Lecture Notes in Computer Science

Edited by G. Goos and J. Hartmanis

179

Victor Pan

How to Multiply Matrices Faster

Springer-Verlag
Berlin Heidelberg New York Tokyo 1984

Author

Victor Pan
State University of New York at Albany
Department of Computer Science
1400 Washington Avenue, Albany, NY 12222, USA

CR Subject Classification (1982): F.2.1, G.1.3

ISBN 3-540-13866-8 Springer-Verlag Berlin Heidelberg New York Tokyo
ISBN 0-387-13866-8 Springer-Verlag New York Heidelberg Berlin Tokyo

Printed in Germany

Printing and binding: Beltz Offsetdruck, Hemsbach / Bergstr.
2146/3140-543210

CONTENTS

Some Notation and Abbreviations

Notation	Meaning, Comments	Defined or First Used In Section(s) (see also Index)
A	algorithm	19,23
ar(A);ar(P)	number of arithmetical operations involved in A; required in order to solve a problem P	19,23
as(A)	number of additions/subtractions involved in A	32,33
AAPR	accumulation of the accelerating power via recursion	6
$b_\gamma(X,Y)$	bilinear form in X,Y	2
BA(n)	bilinear algorithm for nXn MM	2,22,23
BA(n,λ)	bilinear λ-algorithm for nXn MM	23
BBM	Boolean MM	18
bs(A)	bit-space used by A	23
bt(A)	bit-time used by A	23
bt(s),bt(*,s),bt(±,s)		18
bs(P),bt(P)	bit-time and bit-space of a computational problem P	23
C	the field of complex numbers	2
cρ	commutative rank	32
cbρ	commutative λ-rank	33
C(g,h)	g!/(h!(g-h)!)	8,9
cond	condition	25
D	domain of definition of problem or algorithm	Part 2 (Summary); 23
d	degree of λ-algorithm	6
d	shortest distance	18 only

det(W)	determinant of a matrix W	19
Det(n)	the problem of the evaluation of the determinant of an $n \times n$ matrix	Part 2 (Summary); 19
DFT(n)	discrete Fourier transform,	38,39
E	extension of a ring (field)	5
E,E(n),e(n),E(A,D,h), E(Z(V),D,h)	error bounds	Part 2 (Summary); 23-30
F	ring,field	2
$F[\lambda]$	ring of polynomials over F	6
$f(i,j,q)$, $f'(j,k,q)$, $f''(k,i,q)$, $f(\alpha,q)$, $f'(\beta,q)$, $f''(\gamma,q)$	constant coeficients (from F) of bilinear algorithms	2
$f(i,j,q,\lambda)$, $f'(j,k,q,\lambda)$, $f''(k,i,q,\lambda)$, $f(\alpha,q,\lambda)$, $f'(\beta,q,\lambda)$, $f''(\gamma,q,\lambda)$	coefficients (from $F[\lambda]$) of bilinear λ-algorithms	6
$f,f',f'',\tilde{f},\tilde{f}',\tilde{f}''$		23
FFT	fast Fourier transform	Intr.,2,38
h(s)	2^{1-s}	23
u^H, W^H	complex conjugate of number u, conjugate transpose of matrix W	19
I (also I_n))	identity matrix (of size $n \times n$)	19
ℓ_h(ℓ_2, ℓ_∞)	ℓ_h-norm of a matrix or of a vector	24
log u	logarithm to the base 2 of u	1

L_q, L_q'		2
L_q''		10
M	rank of algorithm, λ-rank of λ-algorithm	2,4,6
MA,MS	matrix addition,subtraction	20
MI	matrix inversion	Part 2 (Summary); 19
MM	matrix multiplication	Intr., 1
(m,n,p); also mXnXp MM	the problem of mXn by nXp MM	2
O(g(s)),o(g(s))	see Notation 18.1	Intr.,1,18
O, O_n	null matrix	19,20
PM	polynomial multiplication	2
Q	field of rational numbers	2
Q	unitary matrix (a QR-factor)	20
Q(s)	computed approximation to Q	26-30
$QR, \tilde{Q}R, QR^*$		20
R	upper triangular matrix (a QR-factor)	20
R(s)	computed approximation to R	26-30
R	field of real numbers	2
$\underline{R}$	set of vectors in the proof of Theorem 7.2	9 only
SLE	the problem of solving a system of linear equations	Part 2 (Summary); 19
sm(A)	number of scalar multiplications in A	32,33
T	trilinear form	10
TA	trilinear aggregating	Intr.,3,11

Tr(W)	trace of a matrix W	10
TMI	triangular matrix inversion	21
t	tensor	2,10
U,V,W,X,Y,Z	matrices	1,2,4,6,10
Z	ring of integers	2,5
$Z(\nu)$	ring of integers modulo ν	2,5
Z(V)	output matrix	24-30
$\delta(i,j)$	$\delta(i,j)=0, i \neq j; \delta(i,i)=1$	2
Δ, Δ'	error value,error matrix	23-30
λ	see λ-algorithms	4,6
ρ , ρ_F	rank, rank over a ring F	2
$\rho(m,n,p)$	rank of mXnXp MM	2
$b\rho$	λ-rank	36
ω	exponent of MM	2
ω_F	exponent of MM over a ring F of constants	2
Σ, Π	symbols of sums, products	
Σ	diagonal matrix	20 only
$\lfloor u \rfloor, \lceil u \rceil$	see Notation 18.1	18
$\oplus$	direct sum of disjoint problems	8
$\odot$	direct sum of identical disjoint problems	2,5,8
$\otimes$	(tensor) product of bilinear problems	2,5,8
$\otimes$	direct (Kronecker) product of vectors, matrices, tensors, and of linear, bilinear, or polylinear forms	10,14,16

$\emptyset$	generalized MM	18 only
$\|\|\underline{v}\|\|, \|\|W\|\|$	norms of vector $\underline{v}$, matrix W	24
$t \leftarrow t'$	mapping (algorithm)	5,8
$\|S\|$	cardinality of a set S	
$\|u\|$	absolute value (modulus) of a number u	
$\subset$, $\subseteq$	inclusion of one set into another	5
$\in$	inclusion of an element into a set	9
$\cup$	union of sets	5
■	end of clause, of proof, of algorithm	

Introduction

Matrix multiplication (hereafter referred to as MM) is a basic operation of linear algebra, which has numerous applications to the theory and practice of computation. In particular, several important applications are due to the fact that MM is a substantial part of several successful algorithms for other computational problems of linear algebra and combinatorics, such as the solution of a system of linear equations, matrix inversion, the evaluation of the determinant of a matrix, Boolean MM, and the transitive closure of a graph. Moreover, the computational time required for MM is the dominating part of the total computational time required for all of those problems, that is, all such problems can be reduced to MM and can be solved fast if MM is solved fast.

How fast can we multiply matrices? The product of a pair of $N \times N$ matrices X and Y can be evaluated in a straightforward way using N^3 multiplications of the entries of X and Y and $N^3 - N^2$ additions of the resulting products. The best upper bound on the number of arithmetical operations for MM and for all related problems listed above remained $O(N^3)$ until 1968 while the best lower bounds on that number were of order N^2. The best lower bound is still of the same order N^2 but since 1968 the "upper" exponent 3 has been reduced to the currently record value 2.496. So far the latter progress made no impact on the practice of computing matrix products due to the substantial overhead of the asymptotic acceleration of MM. The impact on the theory of algorithms, however, was substantial. The mere existence of asymptotically fast algorithms for MM and for the related computational problems is encouraging. More important, a deeper insight into the subject of the design of efficient algorithms for arithmetical computations has been obtained due to the study of MM. Furthermore some specific techniques such as the duality method and the methods of the fast approximate computation first passed the tests of their efficiency in the study of MM and then were successfully applied to other computational problems. Also some rather unexpected theoretical applications followed. (For example, see Theorem 7.1 in Section 7, which implies a nontrivial fact of the theory of tensors, and note the quantitative measure of asynchronicity of linear computations defined in Section 39.)

The original intension of the present author was to summarize the knowledge accumulated through the years of the study of MM. In the process of writing the book, the author observed some new facts, developed some new insights, and arrived at some algorithms that promise some practical application.

The major subdivision of the material will be as follows. In Part 1 of this book the algorithms for MM will be devised that are currently asymptotically fastest, that is, they define the exponent 2.496. This will be an occasion to demonstrate all of the major techniques that historically have been applied to the asymptotic acceleration of MM. In Part 2 the computations for several problems of combinatorial analysis and linear algebra will be reduced to MM and consequently accelerated. The traditional reduction in terms of the numbers of arithmetical operations will be extended to the reduction in terms of the numbers of bit-operations and bits of storage space involved. The bit-time and the bit-space used in the algorithms will be shown to be closely related to the values of the condition numbers of those algorithms; those condition numbers will be estimated. The bit-time complexity classification of the linear algebra problems rather unexpectedly differs from their arithmetical complexity classification. In Part 3 several algorithms for fast MM with smaller overhead will be presented. Those algorithms are superior over the asymptotically fastest algorithms for NXN MM for all sizes of interest, say, certainly for all $N < 10^{20}$. Furthermore, the unrestricted λ-algorithms defined in Section 40 promise to lead to an efficient practical method of computation in linear algebra. (That class of algorithms extends our study begun in [P80a] and [P82].) Also in Part 3 some well-known linear lower bounds on the complexity of algorithms of different classes will be derived for the sake of completeness and in order to demonstrate the basic active substitution method and some other customary lower bound techniques. The summaries introducing Parts 2 and 3, the titles of the sections and their first paragraphs may serve as the further guidance to the contents of the book.

To facilitate the selective reading from the book and its use for references, we will minimize the interconnections among Parts 1, 2 and 3 and even among the groups of subjects within each part, particularly within Part 3. (For instance, the readers may now try to read Section 39 on the asynchronicity with the occasional references to the preceding sections, if needed, or they may try to examine Examples 40.1-40.5 of the unrestricted λ-algorithms presented in the concluding Section 40.) On the other hand, we will unify our presentation, at least in Part 1, by using a model example (throughout Part 1) to illustrate the main ideas and the main techniques. (This line has clear historical parallels in the study of MM, which has been greatly influenced by the successful designs of algorithms in the form of modestly looking patterns.) For more selective reading about the asymptotic acceleration of MM, we refer to the papers [P84] (for a lighter presentation of that subject) and [P82c] and [P84a] (for a more concise treatment of it).

We will employ mostly algebraic techniques in Parts 1 and 3 and mostly the techniques of numerical analysis (in particular of error analysis and of numerical linear algebra) in Part 2.

The author tried to make the exposition elementary and basically self-contained. The outside results cited in Part 2 of the book will amount to the well-known estimates for the bit-complexity of the four arithmetical operations and of the evaluation of square roots and to some elementary facts from the undergraduate level text books on numerical analysis. In Parts 1 and 3 a very modest amount of algebra (mostly just the concepts of the algebraic rings and fields) will be used. That amount can be substantially reduced if the readers assume that the computations have been performed with real or complex constants rather than over an arbitrary ring or field of constants and also if they skip the "harder" sections and remarks, such as our remarks on the correlation between the arithmetical complexity and the tensor rank (the concept of tensor rank is not used in our presentation except for the proof of isolated Theorem 36.2 in Section 36 but we indicate where that concept could be used in order to obtain a more comprehensive insight into the subject).

This monograph covers mostly the progress since 1978 that has not been covered in other books so far except for the excellent but very brief treatment in [K]. (Also the papers [P81],[S81], and [CW] remain very far from being complete and updated surveys on MM.) Several adjacent topics have already been well treated in the books and in the survey papers. We will omit those topics or only briefly mention about them referring the reader to [AHU], [BM], [K] on the design and analysis of algorithms for arithmetical computations, to [GvL] on the numerical aspects of linear algebra including the computations with sparse and special matrices; to [BGH] and [He] on parallel algebraic computing; to [DGK] on the pipe-line computation in linear algebra; and to the numerous publications on the computation with Toeplitz, Hankel, and circulant matrices and on the fast Fourier transform (FFT), see, in particular, [AHU], [B83], [BGY], [BM], [F71], [F72], [F72a], [FZ], [FMKL], [GvL], [K], [KKM], [Ra], [W80].

I will end this introduction with a brief overview of the history of MM and with my recollections of my own work on the subject. (In order to emphasize the more personal character of that overview, I will use the singular "I" rather than the plural "we" until the end of the introduction.)

Historically the study of MM was the second of the major subjects that have shaped the modern theory of the algeraic (arithmetical) computational complexity. The pioneering work of 1954 by A.M. Ostrowski, see [Os], introduced the first subject of that theory, that is, the problem of fast evaluation of polynomials; see [K] and [BM] for the history of the research on that subject where I was involved starting with 1959. In 1964 my former Ph.D. thesis advisor in Moscow University, Professor A.G. Vitushkin, suggested me reducing the mentioned earlier gap between the lower bound 2 and the upper bound 3 on the exponent in the case of a system of linear equations. This was a somewhat natural continuation of my thesis work, where I was fortunate to close the gap between the lower and the upper bounds on the number of arithmetical operations required for the evaluation of polynomials, see

[P62],[P64],[P66]. (Actually the basic active substitution techniques from [P62],[P64],[P66] turned out to be also useful for deriving some linear lower bounds on the arithmetical complexity of linear algebra problems, compare [BM], [K], [St72], [AS].) I have soon shifted to the study of MM and noted in December 1965 how NXN MM could be performed in about $N^3/2$ multiplications and in about $3N^3/2$ additions and subtractions, see Example 32.2 in Section 32. That result was a simple and almost straightforward extension of the previously developed techniques for the fast evaluation of polynomials with preconditioning (see [P66] or [K] for surveys) but my presentation of that result in January 1966 at the scientific seminar in Moscow headed by A.S. Kronrod, E.M. Landis, and G.M. Adel'son-Vel'skii was very well received. However, no interest elsewhere was raised, and the result remained unpublished until 1968 when it was rediscovered by S. Winograd, see [W68]. Furthermore, I could not get any support for my study of MM and had to interrupt that study until 1978. (I, however, published a short paper on that subject in 1972, see [P72].) Meanwhile MM became the subject of worldwide interest in 1968, when Volker Strassen discovered that the problem of MM and consequently several related computational problems of linear algebra can be solved in $O(N^{2.808})$ rather than in $O(N^3)$ arithmetical operations. Many scientists understood that discovery as a signal to attack the problem and to push the exponent further down. No further progress, however, followed for about 10 years. It happened that by that time, in 1978, I was able to come back to the study of MM. In September 1977 I started to work at the Department of Mathematics of the IBM Research Center, Yorktown Heights, New York, reporting the results of my work to S. Winograd, whose previous major publications [W67] and [W70] relied on [P66] and who later became the head of that department. In May 1978 I reexamined my algorithmic design published in 1972, improved it, and reduced the exponent first to 2.795 and soon thereafter to 2.781. That progress relied on the special techniques of trilinear aggregating (TA) introduced in my paper [P72] in 1972. TA turned out to be a general efficient method for the design of fast MM algorithms. The power of TA was substantially enhanced by combining it with two other important techniques introduced later (in 1978 and in 1979). Namely, fast algorithms for MM can be derived from the so called any precision aprpoximation algorithms (APA-algorithms) for Disjoint MM. The striking and ingenious idea of using APA-algorithms for the acceleration of MM was first introduced by D. Bini, M. Capovani, G. Lotti and F. Romani in [BCLR]. The efficiency of the application of that idea to MM was substantiated by D. Bini in [B80]. (We will describe the APA-algorithms in this book under the name λ-algorithms.) The idea of using Disjoint MM was due to A. Schönhage, see [S81]. The idea comes quite natural from the analysis of TA and of the so called direct sum problem. (The latter problem was well known in combinatorial analysis and in algebra. For the algebraic computation, that problem was first stated in the form of a conjecture by V. Strassen in 1973, see [St73].) The comparison of the applications of TA in [P72], [P78], [P80] and in the section "Patterns and Exponents" in [S79],[S81] convinces that historically A. Schönhage came to his idea by analyzing TA for the design of APA-algorithms for the so called Partial MM.

Combining the three major techniques (associated with TA, APA-algorithms, and Disjoint MM), the six researchers, D. Bini, M. Capovani, G. Lotti, F. Romani, A. Schönhage, and myself, reduced the exponent to 2.5161 by March 1980. The next "record" reduction to 2.496 was announced soon thereafter (in September 1980) by D. Coppersmith and S. Winograd. Such a reduction was obtained by the recursive application of an earlier pattern of TA originated in [P72]. The recursion itself relied on a special variation of the general techniques of TA. (It is pleasant for me to recall that both authors, that is, D. Coppersmith and S. Winograd, carefully studied the latter techniques from myself in 1978-1979 and in August 1980 while I was working with them at the IBM Research Center. In 1979-1980 they had more chances to continue working on those techniques than myself since the Mathematics Department of the IBM Research Center in Yorktown Heights did not resume the contract with me in September 1979.)

The history of the speed up of MM confirms the crucial importance of nontrivial patterns for the acceleration of MM. Indeed in 1968 MM was accelerated due to Strassen's ingenious pattern for 2X2 MM and then the progress had to wait for ten years (from 1968 to 1978) until the two "modest" constructions from [P72] and [BCLR] introduced the new techniques of TA and the concept of APA-algorithms, respectively, and until the efficiency of those techniques and concept was proven in 1978, see [P78], and in 1979, see [B80], respectively. On the other hand, several results presented in [S79],[S81],[CW] in the form of a well-recognized theory (which could be mistakenly taken as little dependent on [P72],[P78],[BCLR],[B80]) appeared within a short time thereafter.

In spite of some personal disappointment that my study of MM was twice abruptly interrupted at very unappropriate moments, I decided to come back to that subject for a survey of the results at this stage. That idea was first suggested to me by H.S. Wilf and A.C.-C. Yao. The survey eventually turned into this monograph.

While writing the monograph, I reviewed the subject and obtained some new estimates for the bit-complexity of the computations in linear algebra. (This was a continuation of my previous work reported in [P80a], [P81a], [P81c], and [P82].) Those estimates, in particular Theorem 23.7 of Section 23 of this book, helped me to better appreciate the efficiency of the APA-algorithms (λ-algorithms) for the approximate evaluation of arithmetical expressions; finally I decided to define a new extension of the class of APA-algorithms in order to formalize an approach begun in my earlier papers [P80a] and [P82], see Section 40.

Several valuable and important suggestions by A.B. Borodin, J. Hartmanis, and H.S. Wilf helped me to improve the first draft of the monograph.

It is a pleasure to acknowledge the support of my work by the National Science Foundation under the grant MCS 8203232. I very much appreciated (and I hope the readers will also appreciate) the efficiency and quality of publishing work by Springer Verlag and the skill and the patience of Sally Goodall who typed the manuscript using the UNIX typsetting system supervised by my colleague at SUNYA Peter Bloniarz (UNIX is a Trademark of Bell Laboratories).

PART 1. The Exponent of Matrix Multiplication

Summary. The asymptotically fast algorithms for nXn matrix multiplication (MM) involving $O(n^{2.496})$ arithmetical operations are designed with the demonstration of all major techniques historically used for the asymptotic acceleration of MM. The prospects of further progress are briefly discussed in Section 17.

1. The Power of Recursive Algorithms for Matrix Multiplication

The following important algorithm was discovered by Volker Strassen in 1968 (see [St69]).

Algorithm for 2X2 MM. Evaluate the product Q=XY of two given 2X2 matrices

$$X = \begin{bmatrix} x_{00} & x_{01} \\ x_{10} & x_{11} \end{bmatrix} \qquad Y = \begin{bmatrix} y_{00} & y_{01} \\ y_{10} & y_{11} \end{bmatrix}$$

using the following formulae.

$p_0 = (x_{00}+x_{11})(y_{00}+y_{11})$, $p_1 = (x_{10}+x_{11})y_{00}$, $p_2 = x_{00}(y_{01}-y_{11})$,

$p_3 = (-x_{00}+x_{10})(y_{00}+y_{01})$, $p_4 = (x_{00}+x_{01})y_{11}$,

$p_5 = x_{11}(-y_{00}+y_{10})$, $p_6 = (x_{01}-x_{11})(y_{10}+y_{11})$,

$q_{00} = p_0+p_5-p_4+p_6$, $q_{01} = p_2+p_4$, $q_{10} = p_1+p_5$, $q_{11} = p_0-p_1+p_2+p_3$.

For any pair of 2X2 matrices X,Y, this algorithm evaluates the matrix XY in 7 multiplications and 18 additions/subtractions. The straightforward method involves 8 multiplications and 4 additions and seems more efficient. This impression changes, however, if we apply Strassen's algorithm recursively in order to evaluate a $2^h X 2^h$ matrix product.

Indeed, assume that the eight inputs $x_{00},x_{01},\ldots,y_{11}$ of Strassen's algorithm are not just numbers or indeterminates but that they are themselves 2X2 matrices. Then so are $p_0,p_1,\ldots,p_6,q_{00},q_{01},q_{10},q_{11}$. The operations of the algorithm become operations over 2X2 matrices. Using Strassen's algorithm again in order to perform each of the seven 2X2 MM and summing and subtracting matrices in the straightforward way we need perform only $7^2 = 49$ multiplications and $7*18+4*18 = 198$ additions/subtractions. Note that the matrices X,Y,Q become 4X4 matrices, so that we arrive at an algorithm for 4X4 MM. For illustration, here is the block representation of Q (and similarly for X,Y).

$$Q = \left[\begin{array}{cc|cc} q_{00} & q_{01} & q_{02} & q_{03} \\ q_{10} & q_{11} & q_{12} & q_{13} \\ \hline q_{20} & q_{21} & q_{22} & q_{23} \\ q_{30} & q_{31} & q_{32} & q_{33} \end{array}\right] = \begin{bmatrix} Q_{00} & Q_{01} \\ Q_{10} & Q_{11} \end{bmatrix}.$$

The broken lines separate the 2X2 blocks $Q_{00}, Q_{01}, Q_{10}, Q_{11}$ from each other.

Similarly substitute 4X4 matrices for the entries of the original matrices X,Y and continue the above recursive process of constructing algorithms for $2^h X 2^h$ MM, h=3,4,... In this paragraph let M(N) and A(N) designate the numbers of multiplications and additions/subtractions, respectively, required for NXN MM in such recursive algorithms. Then we immediately deduce that $M(2^h) = 7^h, A(1) = 0, A(2^{h+1}) = 18*4^h + 7A(2^h)$, h = 1,2,... Consequently $A(h) = 6(7^h-4^h)$, so that $M(h) + A(h) < 7^{h+1}$ for all h (compare [St69] or, say, [BM], p. 4). The algorithm is extended to NXN MM for all N by padding matrices with zeroes if N is not in the form 2^h. A total of at most $4.7N^{\omega} = O(N^{\omega}), \omega \leq \log 7 = 2.807...$ arithmetical operations suffice for NXN MM. Here the _exponent ω is defined by the number of multiplications in the basis algorithm_. (Additions/subtractions contribute little to the total computational cost because they are substantially simpler than multiplications if we perform all operations over large matrices as we do in this recursive construction. Here and hereafter we apply the customary notation O(s) as $s \to \infty$, compare Notation 18.1 in Section 18, and we assume that _all logarithms are to the base two_.)

Since 2.807...<3, we have _asymptotically accelerated_ the "natural" straightforward method for NXN MM, which uses $2N^3-N^2$ operations. The computation for several other important graph theoretical and linear algebra problems (in particular, for the transitive closure of a graph (see Section 18) and for simultaneous linear equations (see our Sections 19-21)) can be reduced to MM and consequently accelerated, so that the exponent $\omega \leq 2.807...$, soon after Strassen's discovery, became one of the most frequently cited quantities in the study of the time-complexity of algorithms. Strassen's result also raised the question whether the exponent log 7 = 2.807... can be reduced further or, more generally, what is the greatest lower bound ω on the exponents, such that NXN MM can be performed involving a total of $O(N^{\omega+\varepsilon})$ arithmetical operations for arbitrary positive ε? (We cautiously wrote $O(N^{\omega+\varepsilon})$ rather than just $O(N^{\omega})$ allowing ω to be a limiting point rather than just the minimum value.) We will call such ω the _limiting exponent of MM_ or, simply, just the _exponent of MM_.

It is immediately seen that

$$\omega \geq 2. \tag{1.1}$$

Indeed, each arithmetical operation involves two inputs and has one output while each NXN MM problem has $2N^2$ inputs and N^2 outputs. (1.1) remains the best lower bound on the exponent ω that is presently known.

2. Bilinear Algorithms for MM

The development described in the previous section came into being due to the successful design for 2X2 MM in seven multiplications. (Recall that the number of additions/subtractions in that basis algorithm does not influence the exponent.) More generally, if for some pair n, M there exists an algorithm for nXn MM that involves only M multiplications and gives rise to a Strassenlike recursive construction, then

$$\omega \leq \log M/\log n. \tag{2.1}$$

We have n=2, M=7, $\omega \leq \log 7$ for Strassen's algorithm. Next we will define quite a general class of algorithms, which can all be used as the bases of the recursive constructions. As in the case of Strassen's algorithm, we will require that the computational scheme work for all instances of the given input matrices, so that we will consider their entries as independent variables (indeterminates).

Definition 2.1. Bilinear Algorithms for MM. Given a class of constants F and two mXn and nXp matrices $X = [x_{ij}]$, $Y = [y_{jk}]$, then compute XY in the following order. First evaluate the linear forms in the x-variables and in the y-variables,

$$L_q = \sum_{i,j} f(i,j,q)x_{ij}, \quad L'_q = \sum_{j,k} f'(j,k,q)y_{jk}, \tag{2.2}$$

then evaluate the products $P_q = L_q L'_q$ for q=0,1,...,M-1, and finally evaluate the entries $\sum_j x_{ij}y_{jk}$ of XY as the linear combinations

$$\sum_j x_{ij}y_{jk} = \sum_{q=0}^{M-1} f''(k,i,q)L_q L'_q. \tag{2.3}$$

Here f,f',f" are constants from F such that (2.2), (2.3) are identities in the indeterminates x_{ij}, y_{jk} for all i,j,k, i=0,1,...,m-1; j=0,1,...,n-1; k=0,1,...,p-1. M, the total number of all multiplications of L_q by L'_q is called the rank of the algorithm and the multiplications of L_q by L'_q are called the bilinear steps or the bilinear multiplications of the algorithm. All other arithmetical operations in the algorithm, that is, the multiplications by the constants f, f',f'', as well as all additions and subtractions, are called the linear steps of that algorithm or its linear operations. (The straightforward algorithm for mXn by nXp MM is bilinear and has rank mnp. Strassen's algorithm for 2X2 MM is also bilinear and has rank 7. In both straightforward and Strassen's algorithms the constants f,f',f" take only the values 0,1, and (for Strassen's algorithm) -1.) The minimum rank of bilinear algorithms for mXn by nXp MM is called the rank of the problem of mXn by nXp MM and is designated $\rho(m,n,p)$. mXnXp MM as well as the triplet (m,n,p) will designate such a problem of MM itself. We will write nXn MM rather than nXnXn MM.

Hereafter the indices i,j,k,q will range from 0 to m-1, n-1, p-1, M-1, respectively, unless it is indicated otherwise.

Remark 2.1. For example, the fields C,R, or Q of complex, real, or rational numbers, respectively, the ring Z of integers, or the ring $Z(\nu)$ of integers modulo an integer $\nu > 1$ (which is a field for prime ν) can be chosen as the class F of constants in (2.2),(2.3). Furthermore F can be any associative and commutative algebraic ring with the unit element (an algebraic ring is a set closed under additions, subtractions, and multiplications; fields are the commutative rings closed under divisions, see [vdW]). In addition to the rings $C,R,Q,Z,Z(\nu)$, the list of customary rings of constants used in the algorithms for algebraic (arithmetical) computational problems includes the rings of polynomials, polynomials modulo a fixed monic polynomial, formal infinite power series, finite power series, finite power series in λ modulo a positive power of λ, and so on. F should be an algebraic ring with the unit element because we must recursively perform the ring operations (that is, additions, subtractions, and multiplications) starting with the constants f,f',f'' from F and with the input variables x_{ij},y_{jk} (see the right sides of the identities (2.3)) and because finally we should come to the coefficients 0 or 1 of the bilinear forms in those input variables (see the left sides of the identities (2.3)). Therefore F must contain either Z or $Z(\nu)$ for some integer $\nu > 1$. Proving Theorem 5.5 in Section 5 we will show that it is sufficient to deal only with the constants that belong to such a subring Z or $Z(\nu)$ of F or to its extension by divisions and by solving algebraic equations over that subring. (We will ignore the possible modification of (2.2), (2.3) with the permission for the ring F to be noncommutative because we do not know any example of efficient algorithms of that class, compare, however, the end of Section 17.) Hereafter we will mean "associative and commutative rings with the unit element" whenever we say "rings". Strictly speaking, we should (and in some cases will) write $\rho_F(m,n,p)$ and ω_F rather than just $\rho(m,n,p)$ and ω because both ranks and exponent of MM may depend on the choice of F. However, more frequently we will omit the subscript F writing $\rho_F(m,n,p)$ and ω_F only in those relations that do not hold over all rings F. We will assume throughout this book that, in all divisions of the algorithms that we will encounter, the divisors always can be inverted in F where they belong to F themselves. Equivalently we may replace each division by a constant by the multiplications by its inverse.

In this monograph we will focus on bilinear algorithms for MM but actually bilinear algorithms are useful for a more general class of problems, whenever a set of bilinear forms must be evaluated (see [AHU],[BM],[K],[W80]).

Definition 2.1 can be generalized as follows.

Definition 2.2. A bilinear algorithm of rank M for the bilinear problem of the evaluation of a given set of bilinear forms (in the x-variables from a set X and in the y-variables from a set Y)

$$b_\gamma(X,Y) = \sum_{\alpha,\beta} g(\alpha,\beta,\gamma)x_\alpha y_\beta, \quad \gamma = 0,1,\dots,\Gamma-1, \tag{2.4}$$

is defined by the following bilinear identities in the x-variables x_α and in the y-variables y_β for all α,β,

$$\left.\begin{aligned} b_\gamma(X,Y) &= \sum_{q=0}^{M-1} f''(\gamma,q)L_qL'_q, \\ L_q = \sum_\alpha f(\alpha,q)x_\alpha, \quad L'_q &= \sum_\beta f'(\beta,q)y_\beta, \quad q=0,1,\dots,M-1. \end{aligned}\right\} \tag{2.5}$$

The algorithm first evaluates $L_q,L'_q,L_qL'_q$ for $q=0,1,\dots,M-1$, and then $b_\gamma(X,Y)$ for all γ. The M multiplications of L_q by L'_q in (2.5) are called the bilinear steps or the bilinear multiplications of the algorithm (2.5), all other arithmetical operations in the algorithm (2.5), that is, the multiplications by the constants f, f', f'', additions, and subtractions, are called its linear steps or its linear operations. The rank of a bilinear problem is the minimum rank in all bilinear algorithms for that problem.

MM is a specific bilinear problem and (2.2),(2.3) is just a particular case of (2.5) where dealing with 2-dimensional arrays (matrices) we use a pair of subscripts (or indices) (i,j), (j,k), or (k,i) in order to represent each of α,β,γ. In addition, here are two other simple illustrative examples of very well known bilinear problems and algorithms, compare [AHU],[BM],[K],[W80].

Example 2.1. The evaluation of the product of two complex numbers, $(x_0+ix_1)(y_0+iy_1)=(x_0y_0-x_1y_1)+i(x_0y_1+x_1y_0)$, where $i^2=-1$, amounts to the evaluation of the two bilinear forms $x_0y_0-x_1y_1, x_0y_1+x_1y_0$. (Note that here and in the next example we need only one subscript per variable.) The straightforward algorithm is bilinear where $L_0L'_0=x_0y_0, L_1L'_1= x_1y_1, L_2L'_2=x_0y_1, L_3L'_3=x_1y_0, M=4$. Here is a simple bilinear algorithm of rank 3 for the same problem, the constants f,f',f" are also equal to 0,1 and -1. $L_0L_0'=x_0y_1, L_1L_1'=x_1y_0, L_2L_2'=(x_0-x_1)(y_0+y_1)$, $x_0y_0-x_1y_1=-L_0L_0'+L_1L_1'+L_2L_2'$, $x_0y_1+x_1y_0=L_0L_0'+L_1L_1'$.

Example 2.2. Polynomial multiplication, PM(m-1,n-1), PM(d). Given two polynomials in λ,

$$P_{m-1}(\lambda) = \sum_{\alpha=0}^{m-1} x_\alpha \lambda^\alpha, Q_{n-1}(\lambda) = \sum_{\beta=0}^{n-1} y_\beta \lambda^\beta$$

whose coefficients x_α, y_β are considered indeterminates. Evaluate the coefficients of the polynomial $P_{m-1}(\lambda)Q_{n-1}(\lambda)$. This problem (designated by PM(m-1,n-1) and also called the convolution of two vectors $\underline{X} = [x_\alpha]$, $\underline{Y} = [y_\beta]$) is the bilinear problem of the evaluation of the set of bilinear forms $b_\gamma(X,Y) = \sum_{\alpha+\beta=\gamma} x_\alpha y_\beta$, where α, β, and γ range from 0 to m-1, n-1, and m+n-2, respectively. (Here X and Y are just the sets of the entries of the given vectors $\underline{X}$ and $\underline{Y}$, respectively.) We will also consider the subproblem (designated by PM(d)) where it is required to evaluate only the subset $\{b_\gamma(X,Y), \gamma \le d\}$ of the above set such that $d \le m$, $d \le n$. This is the problem of the multiplication of two polynomials in λ modulo λ^{d+1}. The straightforward algorithms for the problems PM(m-1,n-1) and PM(d) have ranks mn and $(d+1)^2$, respectively. The algorithms use only the constants 0 and 1 so that they work over every ring F. If the ring F of constants is a field that contains at least M=m+n-1 different elements $\lambda(0),\ldots,\lambda(M-1)$, then the problem PM(m-1,n-1) can be solved by the following algorithm of rank M, see [T].

Step 1. Evaluate $P_{m-1}(\lambda)$, $Q_{n-1}(\lambda)$ at $\lambda = \lambda(i)$, $i=0,1,\ldots,M-1$.

Step 2. Evaluate the product $P_{m-1}(\lambda)Q_{n-1}(\lambda)$ at $\lambda = \lambda(i)$, $i=0,1,\ldots,M-1$.

Step 3. Find the coefficients of $P_{m-1}(\lambda)Q_{n-1}(\lambda)$ by interpolation.

Steps 1 and 3 require only linear operations. Step 2 involves exactly M bilinear multiplications.

The same algorithm can be applied to the problem PM(d) but the reduction modulo λ^{d+1} of the polynomials $P_{m-1}(\lambda)$, $Q_{n-1}(\lambda)$ and of their product enables us to do with only 2d+1 bilinear multiplications whenever F is a field with at least 2d+1 distinct elements. Our study of the problem PM(d) is immediately extended to the problem of the multiplication of two polynomials in λ modulo a polynomial $\lambda^{d+1} + p_d^*(\lambda)$ where $p_d^*(\lambda)$ is a fixed polynomial of degree at most d. The minimum number of bilinear multiplications remains the same as for PM(d). Note that the multiplication of two complex numbers (Example 2.1) can be considered the multiplication of two polynomials in λ modulo λ^2+1. In such a case we have that d=1, 2d+1=3, so that we come again to reducing one complex multiplication to three real ones.

Finally note that the convolution PM(m-1,n-1) can be equivalently represented as the multiplication of the $(m+n) \times n$ Toeplitz matrix $X = [x_{i-j}, i=0,1,\ldots,m+n-1; j=0,1,\ldots,n-1]$ (where $x_k = 0$ for $k<0$ and $k \geq m$) by the n-dimensional column-vector $\underline{y} = [y_j, j=0,1,\ldots,n-1]$. (The general Toeplitz matrix is obtained by removing the requirements $x_k = 0$.) Similarly the multiplication of two (n-1)-th degree polynomials in λ modulo $\lambda^n - 1$ (called also the cyclic convolution) can be equivalently represented as the multiplication of the $n \times n$ circulant matrix $X = [x_{i-j}, i,k=0,1,\ldots,n-1]$ (where $x_k = x_{k+sn}$ for all integers s, that is, the subscripts are defined modulo n) by the n-dimensional column-vector $\underline{y} = [y_j, j=0,1,\ldots,n-1]$. (Note that circulant matrices are just special Toeplitz matrices.) Due to the reduction back to polynomial multiplication, we may multiply every $m \times n$ Toeplitz matrix and every $n \times n$ circulant matrix by a vector involving only $m+n-1$ and $2n-1$ bilinear steps, respectively, while the straightforward algorithms involve mn and n^2 bilinear steps, respectively. Actually the computation can be performed in fewer linear and bilinear steps (say, a total of $O(n \log n)$ steps suffice if m=n) via the Fast Fourier Transform (FFT), see [AHU], [BM], [F71], [F72a], [FZ], [K] and compare Remark 38.1 in Section 38.

The importance of bilinear algorithms for MM stems from the following theorem.

Theorem 2.1. Given a bilinear algorithm (2.2), (2.3) for the problem (m,n,p), that is, for $m \times n \times p$ MM, such that mnp>1, then

$$\omega \leq 3 \log M/\log(mnp). \tag{2.6}$$

Remark 2.2. Theorem 2.1 reduces estimating the asymptotic complexity of MM (represented by ω) to estimating the rank of a specific problem (m,n,p), so it "only" remains to choose appropriate m,n,p and to design a bilinear algorithm (2.2),(2.3) of a smaller rank M. (V. Strassen did that in [St69] for m=n=p=2.) Note that the exponent depends only on the number of bilinear steps of the algorithm (2.2),(2.3) or equivalently on the number of bilinear terms (products) $L_q L_q'$ in (2.3) and that the number of linear operations involved in the basis bilinear algorithm does not influence the exponent. As can be seen from the proof of Theorem 2.1, see below, the linear operations contribute little to the total cost of the computation because again it is simpler to add/subtract two matrices or to multiply an $N \times N$ matrix by a constant than to multiply it by another $N \times N$ matrix.

At first we will give a proof assuming that m=n=p. We will proceed as in the particular case in the previous section where n=2, M=7. Recursively for h=1,2,... substitute $n^h \times n^h$ matrices for all indeterminates x_{ij}, y_{jk} of the original algorithm. Then L_q, L_q', L_qL_q', and all linear combinations of the x-variables, of the y-variables, and of L_qL_q' in (2.2), (2.3) become $n^h \times n^h$ matrices while X,Y and XY become $n^{h+1} \times n^{h+1}$ matrices. Apply such a new bilinear algorithm $BA(n^{h+1})$, perform all bilinear steps L_qL_q' using the previously defined bilinear algorithm $BA(n^h)$ for $n^h \times n^h$ MM, and perform other matrix operations (that is, all linear steps consisting in the additions and subtractions of matrices and in their multiplications by the constants f,f',f") by the relatively inexpensive straightforward methods. (Such a linear matrix operation involves only n^{2h} arithmetical operations over the entries of the matrices.) It remains to count the arithmetical operations performed at all steps and derive the bound (2.1), which amounts to (2.6) where m=n=p.

Let us supply some estimates. Let again M(N) and A(N) designate the numbers of bilinear multiplications (rank) and of linear operations, respectively, in the above recursive algorithms for NXN MM, so that M(n)=M. Then

$$M(n^{h+1}) = M(n^h)M = M^{h+1}, \; A(n^{h+1}) = A(n)n^{2h} + A(n^h)M =$$

$$A(n)(n^{2h} + Mn^{2(h-1)} + M^2n^{2(h-2)} + \dots + M^h) = A(n)M^h \sum_{g=0}^{h} (n^2/M)^g = \qquad (2.7)$$

$$A(n)M^h(1-(n^2/M)^{h+1})/(1-n^2/M), h=1,2,\dots$$

In the sequel we will see that $M > n^2$. This immediately implies that $M(n^h) + A(n^h) = O(M^h)$ as $h \to \infty$, which leads to the desired bound (2.6) on ω in the case where m=n=p. ∎

The recursive construction of bilinear algorithms that we have just applied can be, of course, generalized to the case of arbitrary m,n,p. Such a construction can be defined by the recursive application of the following basic fact.

<u>Proposition 2.2</u>. Every two bilinear algorithms B^* and B of ranks M^* and M, respectively, for the two problems (m^*,n^*,p^*) and (m,n,p), respectively, define a bilinear algorithm $B^* \otimes B$ of rank at most M^*M for the problem (m^*m,n^*n,p^*p). (The problem (m^*m,n^*n,p^*p) is called the <u>product of the two problems</u> (m^*,n^*,p^*) and (m,n,p) and is designated by $(m^*,n^*,p^*) \otimes (m,n,p)$. The algorithm $B^* \otimes B$ is called the <u>product of the two algorithms</u> B and B^*. The <u>powers</u> $B^{\otimes h} = B^{\otimes(h-1)} \otimes B$ <u>of the algorithm</u> $B = B^{\otimes 1}$ are defined recursively for h=2,3,...)

In the sequel we will represent the bilinear algorithm (2.2),(2.3) (of rank M) by the <u>implication</u> (or, equivalently, by the <u>mapping</u> of one MM problem into another)

$$(m,n,p) \leftarrow M \odot (1,1,1). \tag{2.8}$$

(2.8) can be interpreted as the possibility to solve the problem (m,n,p) by solving M problems $(1,1,1)$ (that is, by performing M bilinear steps of the algorithm (2.2),(2.3)) provided that the linear operations (that is, multiplications by constants, additions, and subtractions) are considered free, compare Theorem 2.1. Then Proposition 2.2 can be restated in the following equivalent form.

Proposition 2.2'. Every two bilinear algorithms B^* and B,

$$(m^*,n^*,p^*) \leftarrow M^* \odot (1,1,1), \ (m,n,p) \leftarrow M \odot (1,1,1), \tag{2.9}$$

define their product $B \otimes B^*$, that is, a bilinear algorithm

$$(m^*m,n^*n,p^*p) \leftarrow M^*M \odot (1,1,1). \tag{2.10}$$

Proof of Proposition 2.2. Apply the given algorithm B^* (of rank M^*) to the problem (m^*,n^*,p^*). Substitute $m \times n$ matrices for the x-variables and $n \times p$ matrices for the y-variables. Apply algorithm B (of rank M) in order to perform all bilinear steps, which become $m \times n$ by $n \times p$ MM. Count the number of bilinear steps in the resulting algorithm $B^* \otimes B$. ∎

Note that the above substitution of matrices for variables can be considered the multiplication of the two problems (m^*,n^*,p^*) and (m,n,p) that defines the problem (m^*m,n^*n,p^*p) as their product,

$$(m^*m,n^*n,p^*p) = (m^*,n^*,p^*) \otimes (m,n,p). \tag{2.11}$$

Similarly the problem $M^*M \odot (1,1,1)$ can be considered the product of the two problems $M^* \odot (1,1,1)$ and $M \odot (1,1,1)$. Then the product (2.10) of the two algorithms (2.9) can be defined if we simultaneously multiply together the left sides of the two mappings (2.9) as well as their right sides.

Remark 2.3. The constants f,f',f'' of the algorithm $B^* \otimes B$ can be simply expressed through the constants of the algorithms B^* and B. Namely, let (as in (2.2)) the pairs of subscripts be assigned to the variables of the problems (m^*,n^*,p^*) and (m,n,p), that is, designate the variables $x^*_{i^*j^*}$, $y^*_{i^*j^*}$ and x_{ij}, y_{jk}, respectively. Let the variables of the product (m^*m,n^*n,p^*p) of those two problems be represented with 4-tuples of indices as follows: $x(i^*,i,j^*,j)$, $y(j^*,j,k^*,k)$. (To simplify the notation, we have raised the subscripts.) Similarly, let (as in the equations (2.2),(2.3)) the triplets of natural numbers be used in order to define the constants of the algorithms B^* and B, that is, $f^*(i^*,j^*,q^*)$, $f^{*'}(j^*,k^*,q^*)$, $f^{*''}(k^*,i^*,q^*)$, $f(i,j,q)$, $f'(j,k,q)$, $f''(k,i,q)$. Then the constants of the product of those two algorithms $B^* \otimes B$ can be represented with the 6-tuples of indices as follows,

$$\left.\begin{aligned} f(i^*,i,j^*,j,q^*,q) &= f(i^*,j^*,q^*)f(i,j,q), \\ f'(j^*,j,k^*,k,q^*,q) &= f^{*'}(j^*,k^*,q^*)f'(j,k,q), \\ f''(k^*,k,i^*,i,q^*,q) &= f^{*''}(k^*,i^*,q^*)f''(k,i,q) \end{aligned}\right\} \quad (2.12)$$

for all i^*,i,j^*,j,k^*,k,q^*,q. The equations (2.12) can be easily verified directly (see also Proposition 10.1 and Remark 10.2 in Section 10 for a unified proof of (2.12) and of some other basic facts about the recursive algorithms).

Remark 2.4. Those readers who are familiar with the concept of tensors have probably already recognized that every bilinear problem (2.4) can be represented by the 3-dimensional tensor (array) of the coefficients $g(\alpha,\beta,\gamma)$ for all α,β,γ. The rank of the tensor is customarily defined precisely in the same way as the rank of the associated bilinear problem is defined in Definition 2.2, see [Bou], A2, 111-112; [St73]. We do not assume the reader's familiarity with tensors and refer to them only for illustration. However, those readers who represent algorithms by tensors may have some benefits of the unified point of view on the recursive construction of bilinear algorithms, see Remark 8.2 in Section 8. The tensorial representation could also help in the proof of Proposition 16.5 in Section 16. Thus we will give some comments on that interpretation in our remarks here and in the sequel. In particular in the case of MM the associated tensor becomes the 6-dimensional array $t = \{\delta(g,h)\delta(r,s)\delta(u,v), g,h=0,1,\dots,m-1;\ r,s=0,1,\dots,n-1, u,v=0,1,\dots,p-1\}$ because a pair of indices stands for each of α,β,γ. Here and hereafter

$$\delta(i,j) = 0 \text{ if } i \neq j,\ \delta(i,i) = 1.$$

Representing the problems (m,n,p), (m^*,n^*,p^*), $(1,1,1)$, $M \odot (1,1,1)$, $M^* \odot (1,1,1)$ by the associated tensors we may interpret the algorithms (2.9),(2.10) as the mappings of tensors. The notation $\otimes$ and $\odot$ becomes the familiar notation for the products of two tensors and of a scalar and a tensor, respectively. In particular, the substitution of matrices for the variables in the proof of Proposition 2.2 can be interpreted as the multiplication of both sides of the first mapping of (2.9) by the tensor of the coefficients of the problem (m,n,p) or, equivalently as the multiplication of the first mapping of (2.9) by the trivial mapping $(m,n,p) \leftarrow (m,n,p)$ with the subsequent substitutions of the equation (2.11) on the left side of the resulting mapping and of the similar equation $(1,1,1) \otimes (m,n,p) = (m,n,p)$ on the right side. Then the substitution of the second mapping of (2.9) on the right side immediately leads to (2.10). We may interpret Proposition 2.2' as a fact about the construction of the tensor product (2.10) of the two mappings (2.9) (compare Proposition 6.2 in Section 6 and Proposition 8.1 in Section 8). Then the <u>construction of recursive algorithms becomes equivalent to the tensor product construction</u>. In particular, the tensor of the coefficients of the product of the two bilinear algorithms can be expressed as the product of the two tensors of the coefficients of the two original algorithms

$$\{f^*, f^{*'}, f^{*''}\} \otimes \{f, f', f''\} = \{f^* f, f^{*'} f', f^{*''} f''\}.$$

This is an equivalent version of (2.12), compare Remark 10.2 in Section 10.

In order to complete the proof of Theorem 2.1 for all m,n,p, we need the following <u>duality theorem</u> (due to [P72], compare also [HM]) to be proven in Section 10.

<u>Theorem 2.3</u>. For all natural numbers m,n,p,M and for every algorithm of rank M for the problem (m,n,p), it is possible to generate five dual algorithms of the same rank M for the five dual problems (n,p,m), (p,m,n), (m,p,n), (n,m,p), (p,n,m), so that $\rho(m,n,p) = \rho(n,p,m) = \rho(p,m,n) = \rho(m,p,n) = \rho(n,m,p) = \rho(p,n,m)$.

Now Theorem 2.1 for all m,n,p is easily reduced to the case where m=n=p (square MM). Indeed, given a bilinear algorithm B of rank M for (m,n,p), then apply Theorem 2.3 and derive the two dual algorithms B^* and B^{**} of the same rank M for the problems (n,p,m) and (p,m,n), respectively. Apply Proposition 2.2 twice and define the product $(B \otimes B^*) \otimes B^{**}$, which is an algorithm of rank M^3 for the problem $(m,n,p) \otimes (n,p,m)) \otimes (p,m,n) = (mnp,npm,pmn)$. The latter problem is a problem of a square MM where we can already apply Theorem 2.1. Hence $\omega \leq \log M^3/\log(mnp) = 3 \log M/\log(mnp)$, compare (2.6). ∎

The following result can be considered a converse of Theorem 2.1. It shows that, whenever an arbitrary algorithm defines a bound on the exponent of MM, the same bounds can be defined by Theorem 2.1 applied to some bilinear algorithms for MM. (We will not need this result for the design of our fast algorithms and we will postpone its proof until Section 34, see Remark 34.1.)

Theorem 2.4. For arbitrary positive ε there exists $n=n(\varepsilon)$ (which can be chosen arbitrarily large) such that

$$\omega + \varepsilon > \log \rho(n,n,n)/\log n,$$

Equivalently (see Theorem 2.1) ω equals the greatest lower bound in n on $\log \rho(n,n,n)/\log n$. Here the ranks $\rho(n,n,n)$ and the exponent ω of MM are defined over the same (but arbitrary) ring or field F of constants.

3. The Search for a Basis Algorithm and the History of the Asymptotic Acceleration of MM.

Theorem 2.1 suggests that in the vast variety of all possible bilinear algorithms (2.2), (2.3) for all m,n,p, M we need to find one such that

$$3 \log M/\log(mnp) < \log 7.$$

The desired basis algorithms that would satisfy the latter inequality seemed to be quite carefully hidden from the researchers for about 10 years after Strassen's discovery in 1968. At first it was proven that M=6 is not possible for 2X2 MM (see [HK69],[HK71],[W71]). Then efforts were concentrated on the reduction of the rank of algorithms for 3X3 MM to 21 because $\log 21/\log 3=2.771...$ while $\log 22/\log 3=2.813...$ (greater than $\log 7$). However, so far only M=23 has been achieved (see [L]).

Progress resumed only in 1978 with the design where m=n=p=70, M=143640, which yielded $\omega \leq \log 143640/\log 70=2.795...$ Thereafter several new improvements followed as is indicated in the diagram at the end of this section, see also our history sketch in the Introduction.

Actually the numbers n=70, M=143640 came from a more general construction where bilinear algorithms for nXn MM were first defined for all even n such that $M = M(n) = (n^3 - 4n)/3 + 6n^2$. Then it remained to choose the value of n that minimizes $\log((n^3-4n)/3+6n^2)/\log n$ and this turns out to be n=70. Note that our formula would imply only the trivial exponent if we choose $n \leq 8$.

The latter construction and ultimately all so far known asymptotically fastest algorithms that imply $\omega < 2.5$ have originated in a simpler design of [P72] where bilinear algorithms for nXn MM have been presented for all even n such that $M=0.5n^3+3n^2$. That design remained unnoticed by most of the researchers until it was improved in 1978, see [P78], because the application of (2.1) to the latter formula yields at best an exponent slightly below 2.85 (for n=34). Nevertheless, we will be able to demonstrate all of the most important techniques for the asymptotic acceleration of MM by modifying and improving the latter design. We will present that design in a modified version as the bilinear algorithm (4.1),(4.2) below although originally it appeared in the equivalent trilinear version and was applied to the evaluation of one rather than two matrix products. We will end this section by representing a diagram of the progress in the reduction of the exponent.

Diagram. ($\omega(\tau)$ is the best exponent announced by time τ.)

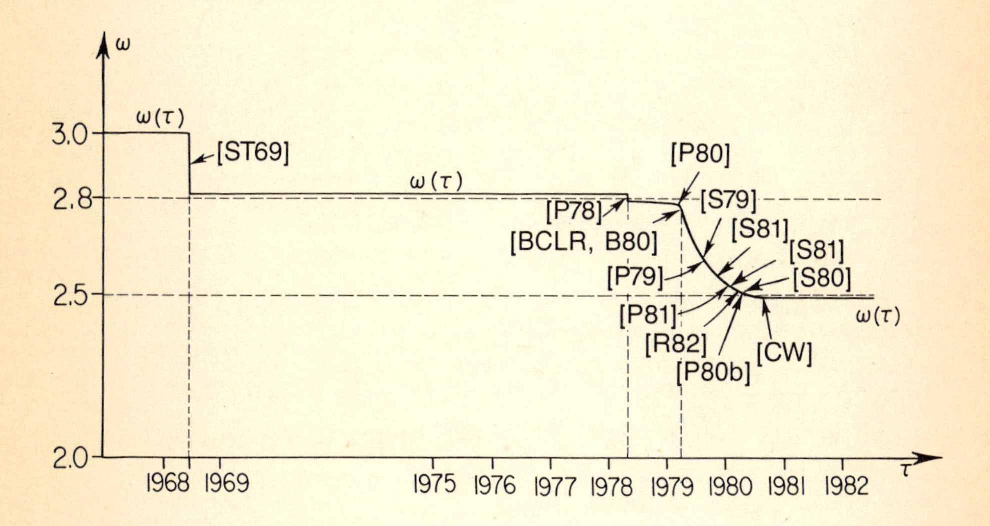

In the diagram the arrows indicate the time of the announcement of the new exponents. For instance, the exponents 2.5167, 2.5161, and 2.496 were announced during 1980 soon after each other; the exponents 2.548 and 2.522 were announced at the Symposium 1979 on the Complexity Theory in Oberwolfach, West Germany, on October 24 and 26, respectively. These facts and the whole diagram demonstrate how extremely rapid the progress was during 1978-1980. It was not clear at that time how the different successful techniques of devising fast algorithms for matrix multiplication were related to each other. In some cases the substantial reduction of the exponents was achieved merely by appropriate combinations of the already available different techniques. For these reasons, the history of fast MM is not fully represented by the dates of the publication and even by the dates of the announcement of the new exponents, so that we also refer the reader to our overview of the subject in the introduction and to some additional comments in the sequel, see, in particular, Remark 6.2 in Section 6 and Remark 13.1 in Section 13.

4. The Basic Algorithm and the Exponent 2.67.

In this section we will outline how to reduce the exponent below 2.67. Our sketchy arguments will be substantiated in the next sections. We will start with the following bilinear algorithm for the simultaneous evaluation of two matrix products XY and UV.

$$\sum_j x_{ij}y_{jk} = \sum_j (x_{ij}+u_{jk})(y_{jk}+v_{ki}) - \sum_j u_{jk}y_{jk} - \sum_j (x_{ij}+u_{jk})v_{ki} \text{ for all } k,i. \quad (4.1)$$

$$\sum_k u_{jk}v_{ki} = \sum_k (x_{ij}+u_{jk})(y_{jk}+v_{ki}) - \sum_k u_{jk}y_{jk} - x_{ij}\sum_k (y_{jk}+v_{ki}) \text{ for all } i,j. \quad (4.2)$$

In this case we assume that for the four given matrices

$$X=[x_{ij}],\ Y=[y_{jk}],\ U=[u_{jk}],\ V=[v_{ki}],\ i=0,\dots,m-1; j=0,\dots,n-1; k=0,\dots,p-1,$$

of the sizes mXn, nXp, nXp, pXm, respectively, we evaluate the mnp bilinear products $(x_{ij}+u_{jk})(y_{jk}+v_{ki})$ for all i,j,k; the mn products $x_{ij}\sum_k(y_{jk}+v_{ki})$ for all i,j; the np products $u_{jk}y_{jk}$ for all j,k , and the pm products $\sum_j(x_{ij}+u_{jk})v_{ki}$ for all k,i. This amounts to a total of

$$M_{m,n,p} = mnp+mn+np+pm \quad (4.3)$$

bilinear products of the form L_qL_q', so that $M_{m,n,p}$ is the rank of the bilinear algorithm (4.1),(4.2). When all of the $M_{m,n,p}$ products L_qL_q' have been found, we can use the identities (4.1),(4.2) in order to evaluate the matrix products XY and UV as the linear combinations of those L_qL_q' with the coefficients f'' equal to 0,1, and -1 and defined by (4.1),(4.2).

We assume no relations among the entries of the four given matrices X,Y,U,V, so we will consider those entries indeterminates and will call XY and UV two <u>disjoint matrix products</u>. The evaluation of two or several disjoint matrix products will be called <u>Disjoint MM</u>.

In Sections 6-9 we will generalize Theorem 2.1 and will see how a bilinear algorithm for Disjoint MM also gives rise to a recursive construction and how this bounds the exponent. In particular the following bound will be deduced from the algorithm (4.1),(4.2),

$$\omega \leq 3\log(M_{m,n,p}/2)/\log(mnp). \tag{4.4}$$

Comparing the latter bound with (2.6) we can see that the factor 2 appears because the algorithm (4.1),(4.2) simultaneously solves two equally hard problems (m,n,p) and (n,p,m), compare Theorem 2.3.

In order to accentuate the power of the algorithm (4.1),(4.2), we will modify it further by introducing an auxiliary nonzero parameter λ as follows.

$$\sum_j x_{ij}y_{jk} = \sum_j (x_{ij}+u_{jk})(y_{jk}+\lambda v_{ki}) - \sum_j u_{jk}y_{jk} - \lambda\sum_j (x_{ij}+u_{jk})v_{ki}, \tag{4.5}$$

$$\sum_k u_{jk}v_{ki} = \lambda^{-1}\{\sum_k (x_{ij}+u_{jk})(y_{jk}+\lambda v_{ki}) - \sum_k u_{jk}y_{jk} - x_{ij}\sum_k (y_{jk}+\lambda v_{ki})\}. \tag{4.6}$$

For every value $\lambda \neq 0$ the algorithm (4.5),(4.6) defines a bilinear algorithm of rank $M_{m,n,p}$ that evaluates XY and UV. On the other hand, consider the case where λ is a real or complex parameter that converges to zero and note that

$$\lim_{\lambda \to 0} \lambda \sum_j (x_{ij}+u_{jk})v_{ki} = 0.$$

Therefore if we delete the "vanishing" terms (products) $-\lambda\sum_j (x_{ij}+u_{jk})v_{ki}$ from (4.5), then we will have a bilinear algorithm of rank

$$M_{m,n,p,\lambda} = mnp+mn+np. \tag{4.7}$$

That algorithm approximately evaluates XY and UV with a precision that can be made arbitrarily small by choosing λ small. This is an example of a <u>bilinear λ-algorithm</u> or equivalently of an <u>arbitrary precision approximation (APA) bilinear algorithm</u>. The class of APA-algorithms (λ-algorithms) for MM was first introduced in [BCLR] where the following nontrivial λ-algorithm was presented.

<u>Example 4.1</u>.

$p_1=(\lambda x_{01}+x_{10})(\lambda y_{01}+y_{10})$, $p_2=(\lambda x_{00}-x_{10})(y_{00}+\lambda y_{01})$,

$p_3=(-\lambda x_{01}+x_{11})(y_{10}+\lambda y_{11})$, $p_4=x_{10}(y_{00}-y_{10})$, $p_5=(x_{10}+x_{11})y_{10}$,

$q_{00}=x_{00}y_{00}+x_{01}y_{10}=\lambda^{-1}(p_1+p_2+p_4)-\lambda(x_{00}+x_{01})y_{01}$,

$q_{10}=x_{10}y_{00}+x_{11}y_{10}=p_4+p_5$, $q_{11}=x_{10}y_{01}+x_{11}y_{11}=\lambda^{-1}(p_1+p_3-p_5)-\lambda x_{01}(y_{01}-y_{11})$.

This is a λ-algorithm for Partial MM, because it evaluates three (out of four) entries of the 2X2 matrix Q=XY (with arbitrary precision as λ decreases). Ignoring the vanishing terms we reduce the rank to 5 while it is possible to prove that the rank of any conventional bilinear algorithm for the same problem equals at least 6, compare the proof of (35.1) in Section 35. Now consider the problem of the evaluation of the product Q^* of a 2X2 matrix X^* by a 2X3 matrix Y^*. Represent Q^* as the sum $Q^*=Q^0+Q^1$ where Q^0, Q^1 take the forms

$$Q^0 = \begin{bmatrix} q^*_{00} & 0 & 0 \\ q^*_{10} & q^*_{11} & 0 \end{bmatrix}, \quad Q^1 = \begin{bmatrix} 0 & q^*_{01} & q^*_{02} \\ 0 & 0 & q^*_{12} \end{bmatrix}.$$

Then evaluate Q^0 and the transpose of Q^1 by the above algorithm. This defines a bilinear λ-algorithm of rank 10 for Q^* and consequently defines the exponent 2.779... It is not the best exponent that can be obtained from this design. It is possible to devise a clever and complicated recursive construction that starts directly from the above design of rank 5 for Partial MM and finally derive the exponent $\omega \leq 3 \log_6 5 = 2.694...$ (see [S79],[S81]). ∎

Rather than delving into the latter approach we will generate a recursive construction starting from our λ-algorithm (4.5),(4.6) that defines the exponent

$$\omega \leq 3 \log(M_{m,n,p,\lambda}/2)/\log(mnp). \tag{4.8}$$

In order to obtain the best numerical value for the exponent ω from (4.8), we substitute (4.7) into (4.8) for m=p=7, n=1 and deduce that

$$\omega \leq 3 \log 31.5/\log 49 = 2.669... \tag{4.9}$$

In the sequel we will show that <u>the latter inequality bounds the number of arithmetical operations</u> not only in the best approximation algorithms for MM but <u>also in the conventional algorithms</u> where the matrix products are evaluated exactly.

The same design (4.5),(4.6) (if appropriately modified and recursively applied) implies the bound $\omega < 2.5$ (see Section 16). However, our next objective is to justify the bounds (4.8),(4.9) as a consequence of the identities (4.5),(4.6). This will also be an occasion to demonstrate some basic algebraic techniques customarily used in the study of algorithms for algebraic computation. Those readers who are interested only in the design of fast algorithms for MM may skip Sections 5-9 (where we prove the inequalities (4.8),(4.9) and also extend Theorem 2.1 so that a certain exponent of MM follows from every bilinear algorithm for Disjoint MM, such as the algorithms (4.1),(4.2) and (4.5),(4.6), see Theorem 7.2 in Section 7). Such readers may go directly to Sections 10-16 where we give a general approach to the design of the desired basis algorithms for Disjoint MM.

5. The dependence of the Exponent of MM on the Class of Constants Used.

In this section we are going to study the dependence of the exponent of MM on the class F of constants. Those readers who are not interested in that subject may skip this section because it has almost no direct correlation with the subsequent ones except for the applications of Proposition 5.2, whose proof is, however, repeated in the next section. On the other hand, this section might be a helpful introduction to the next one, due to the similarity of some techniques used. As our first illustration of those techniques, we will prove the following simple fact.

Proposition 5.1. The exponents of MM over the fields C and R of real and complex numbers coincide with each other, $\omega_R = \omega_C$.

Proof. At first note that each bilinear multiplication $L_q L_q'$ over C can be reduced to at most three bilinear multiplications over the real and imaginary parts of L_q and L_q' (see Example 2.1). Therefore

$$\rho_R(N,N,N) \leq 3\rho_C(N,N,N) \text{ for all } N. \tag{5.1}$$

It remains to observe that the factor 3 little influences the exponents defined by (2.1) provided that $M = \rho(N,N,N)$ and $n=N$ is very large. For a formal proof, we may proceed as follows.

For arbitrary positive ε choose a natural number $N = N(\varepsilon)$ sufficiently large such that

$$\log 3/\log N \leq \varepsilon, \ \log \rho_C(N,N,N)/\log N \leq \omega_C + \varepsilon, \tag{5.2}$$

compare Theorem 2.4. Then apply Theorem 2.1 and derive that

$$\omega_R \leq \log \rho_R(N,N,N)/\log N. \tag{5.3}$$

Combine (5.1)-(5.3) together. Deduce that $\omega_R \leq \log(3\rho_C(N,N,N))/\log N \leq \log \rho_C(N,N,N)/\log N + \log 3/\log N \leq \omega_C + 2\varepsilon$. Choose $\varepsilon \to 0$ and derive that $\omega_R \leq \omega_C$. Since $R \subset C$, it follows that also $\omega_R \geq \omega_C$. ∎

The proof of Proposition 5.1 suggests that $\omega_F = \omega_E$ whenever simultaneously $F \subset E$ and every bilinear multiplication in E can be simulated by h bilinear multiplications in F for some constant h. The following definition introduces the class of rings where the latter property is easy to verify.

Definition 5.1, compare [vdW]. Let F be a ring. Let E be the ring of all polynomials over F modulo a fixed monic polynomial over F of degree d+1 for a positive d. (The elements of E can be identified with polynomials of degree at most d over F.) Then E is called a *simple algebraic extension of F of degree* d. (Note that $F \subset E$.) If $E_0 \subset E_1 \subset E_2 \subset \ldots$ is a sequence of rings such that E_{i+1} is a simple algebraic extension of E_i for every $i \geq 0$, then E_k is called a *finite algebraic extension* of E_0 and $\bigcup_{i=0}^{\infty} E_i$ is called an *(infinite) algebraic extension* of E_0.

Example 5.1. The field C is an example of a simple algebraic extension of a field R. That extension has degree 1 and is defined by the monic polynomial λ^2+1 irreducible over R.

Example 5.2. Every monic polynomial of degree d+1, where d>0, over the ring of integers defines a simple algebraic extension of that ring.

Example 5.3. Every infinite sequence of monic polynomials in λ with rational coefficients whose degrees are monotone increasing defines an infinite algebraic extension of the field of rational numbers.

Next we will easily extend the inequality (5.1) as follows.

Proposition 5.2. If E is a simple algebraic extension of degree d of a ring F defined by a monic polynomial $p(\lambda)$ of degree d+1, then $\rho_F(N,N,N) \leq (d+1)^2\rho_E(N,N,N)$ for all N. The factor $(d+1)^2$ can be replaced by the factor 2d+1 if F is a field with at least 2d+1 distinct elements.

The *proof* of Proposition 5.2 extends the proof of the inequality (5.1) in the same way as Example 2.2 extends Example 2.1. Consider every operation of a bilinear algorithm of rank $\rho_E(N,N,N)$ as an operation over polynomials in λ modulo $p(\lambda)$. Reduce such an operation to operations with coefficients of those polynomials. Note that every linear operation, that is, addition, subtraction, or multiplication by a constant from F, requires only linear operations in F, while every bilinear step over E becomes a multiplication of two polynomials over F modulo $p(\lambda)$. Using the straightforward bilinear algorithm, we will reduce such a bilinear step to at most $(d+1)^2$ bilinear multiplications in F and to some linear operations in F, compare Example 2.2. Summarizing we arrive at a bilinear algorithm in F for (N,N,N) whose rank is at most $(d+1)^2$ times greater than the rank of the original algorithm over E. Performing the bilinear steps over E by the interpolation algorithm of Example 2.2 we may reduce the factor $(d+1)^2$ to 2d+1 if F is a field with at least 2d+1 elements. ■

Applying Proposition 5.2 recursively we extend it as follows.

Proposition 5.3. If E is a finite algebraic extension of a ring F, then $\rho_F(N,N,N) \leq D\rho_E(N,N,N)$ for all N and for some invariant in N constant D.

Remark 5.1. Propositions 5.2 and 5.3 can be immediately generalized to the case of the ranks of bilinear algorithms for all bilinear problems over a ring F and over its simple and its finite algebraic extensions E, respectively.

Next we will easily extend Proposition 5.1 as follows.

Proposition 5.4. If E is a finite algebraic extension of a ring F, then $\omega_E = \omega_F$. (The proof is similar to the proof of Proposition 5.1 but Proposition 5.3 should be applied at the place of the inequality (5.1).)

Next we will much further narrow the dependence of the exponents of MM on the rings of constants used. We will need the following definitions.

Definition 5.2. The minimum ring that contains a given ring F and the roots r of all linear equations of the form fr=1, where f is an element of F that does not divide zero, is called the rational closure of the ring F and is designated F(/). If a ring F coincide with its rational closure, then it is called a ring closed under division. (If Q(F) is the rational closure of F, then it is easy to check that Q(F) is also the rational closure of itself. In particular all fields are closed under division, that is, each of them is the rational closure of itself so that Q(/)=Q, R(/)=R, C(/)=C.)

Example 5.4. The field Q of all rational numbers is the rational closure of the ring Z of integers.

Example 5.5. Every finite ring F is closed under division, that is, it is the rational closure of itself. (Indeed, let f be an element of F. Then, for some positive d and h, $f^d = f^{d+h}$ since F is a finite ring. Therefore, $f^d(f^h-1) = 0$, $f^h - 1 = 0$, provided that f does not divide 0. Hence $f*f^{h-1} = 1$. ■) In particular the ring $Z(\nu)$ of integers modulo ν is closed under division for every ν.

Definition 5.3. The minimum subring of an associative and commutative ring F (with unity) is designated F_1. Its rational closure is designated $F_1(/)$.

It is easy to classify the minimum subrings of all rings and the rational closures of those subrings using the following concept.

Definition 5.4. The characteristic $\nu = \nu(F)$ of a ring F is either the minimum natural $\nu > 0$ such that $\nu * 1 = 0$ in F or ∞ if $\nu * 1 \neq 0$ for all natural $\nu > 0$.

It is immediately verified that

$$F_1 = Z,\ F_1(/) = Q \text{ if F has characteristic } \infty , \tag{5.4}$$

$$F_1 = F_1(/) = Z(\nu) \text{ if F has characteristic } \nu \neq \infty , \tag{5.5}$$

compare Example 5.5.

Theorem 5.5, compare [P81],[S79],[S81]. For all associative and commutative rings F (with unity),

$$\omega_{F_1(/)} = \omega_{F(/)} \leq \omega_F \leq \omega_{F_1}.$$

Before proving Theorem 5.5, we will state the following immediate corollary from that theorem and from (5.4) and (5.5).

Corollary 5.6. Let F be an associative and commutative ring of characteristic $\nu = \nu(F)$. If ν is finite, then $\omega_F = \omega_{Z(\nu)}$. Otherwise $\omega_Q \leq \omega_F \leq \omega_Z$ and furthermore $\omega_F = \omega_Q$ if $F=F(/)$.

Corollary 5.6 implies that ω_F depends only on the characteristic $\nu = \nu(F)$ unless ν is infinite and F is not closed under division.

Corollary 5.6 leads to the following open problems: prove or disprove that (i) $\omega_Q = \omega_Z$; (ii) $\omega_Z = \omega_{Z(\nu)}$ for all ν. (It is known that $\omega_Q \leq \omega_Z$ and $\omega_{Z(\nu)} \leq \omega_Z$ for all ν because

$$\rho_Q(m,n,p) \leq \rho_Z(m,n,p),\ \rho_{Z(\nu)}(m,n,p) \leq \rho_Z(m,n,p) \text{ for all } m,n,p,\nu. \tag{5.6}$$

The former inequality of (5.6) follows because $Z \subset Q$. The latter inequality is derived by the reduction of the bilinear identities (2.2),(2.3) modulo ν.)

Proof of Theorem 5.5. As follows from Theorem 2.4, for arbitrary positive ε and for all sufficiently large n

$$\log \rho_F(n,n,n) / \log n \leq \omega_F + \varepsilon . \qquad (5.7)$$

Rewrite the identities (2.2),(2.3) as the following system of n^6 cubic equations in the $3n^2M$ variables f(i,j,q), f'(g,h,q),f''(r,s,q), where i,j,g,h,r,s range from 0 to n-1.

$$\sum_{q=0}^{M-1} f(i,j,q)\, f'(g,h,q)\, f''(r,s,q) = \delta(j,g)\delta(h,r)\delta(s,i). \qquad (5.8)$$

Here $\delta(u,v)$ equals 0 if $u \neq v$ and equals 1 otherwise.

By the definition of the rank of the problem of n X n MM over F, the system (5.8) has a solution in F for $M = \rho_F(n,n,n)$. On the other hand, (5.8) is a system of algebraic (polynomial) equations with the coefficients equal to 0 and 1. We may obtain a solution to (5.8) by applying the well-known techniques of algebraic elimination to that system (see [vdW]). The latter techniques require only the field operations, that is, additions, subtractions, multiplications, and divisions, and a finite algebraic extension of the minimum subring F_1 of F. Thus the system (5.8) has a solution for $M = \rho_E(n,n,n)$ over a finite algeraic extension E of $F_1(/)$. Therefore, combining (5.7) with Theorem 2.1 and Proposition 5.4, we obtain that $\omega_{F_1(/)} = \omega_E \leq \log M/\log n \leq \omega_F + \varepsilon$ for all positive ε. Consequently $\omega_{F_1(/)} \leq \omega_F$. Other relations of Theorem 5.5 immediately follow because $F_1(/) \subseteq F(/)$, $F_1 \subseteq F \subseteq F(/)$. ∎

Remark 5.2. The results of this section are of rather theoretical nature because the factors $3,(d+1)^2$, 2d+1, and D in the inequality (5.1) and in Propositions 5.2,5.3, respectively, should not be ignored in the analysis of the complexity of any practical algorithm for MM. Thus we are still left with the practical problem of designing fast MM algorithms over different rings of constants. On the other hand, the main tool of our analysis in this and in the next sections, that is, the application of the back and forth transition from the computation in a ring to the computation in its algebraic extension, turned out to be helpful also for the solution of some other algebraic computational problems, which are important in some practical applications, see [W80].

6. λ-algorithms and Their Application to MM. Accumulation of the Accelerating Power of λ-algorithms Via Recursion.

In this section we will justify the deletion of the vanishing terms of the λ-algorithm (4.5),(4.6) as our first step of proving the bounds (4.8),(4.9). We will achieve that goal by purely algebraic methods, which have nothing to do with the approximation by setting $|\lambda|$ to be small. The final fast algorithms will be λ-free bilinear algorithms and will exactly evaluate the matrix products (the application of λ-algorithms for the approximate evaluation of matrix products will be revisited in Section 23, see Theorem 23.7).

At first we will define the general class of bilinear λ-algorithms that includes the λ-algorithms of Section 4 (see (4.5),(4.6) and Example 4.1) as its particular instances. Then we will show that every λ-algorithm can be turned into a conventional bilinear algorithm over some algebraic extension ring of constants. Then we will extend Proposition 2.2 in order to define recursive bilinear λ-algorithms. Finally we will apply Proposition 5.2 and extend Theorem 2.1 to the case of λ-algorithms.

Definition 6.1, compare Definition 2.2. A bilinear λ-algorithm of λ-rank M and of nonnegative degree d for a bilinear problem of the evaluation of a given set of bilinear forms (2.4) over a ring F is defined by the following bilinear identities in λ, in x_α for all α, and in y_β for all β.

$$b_\gamma(X,Y) = \lambda^{-d} \sum_{q=0}^{M-1} f''(\gamma,q,\lambda)\, L_q L_q' + \lambda P \text{ for all } \gamma \tag{6.1}$$

where

$$L_q = \sum_{\alpha} f(\alpha,q,\lambda) x_\alpha,\ L_q' = \sum_{\beta} f'(\beta,q,\lambda) y_\beta \text{ for all } q. \tag{6.2}$$

Here P is a polynomial in the x-variables, in the y-variables, and in λ with coefficients from F and f, f', f'' are polynomials of degree at most d in λ with coefficients from F. The minimum λ-rank in all bilinear λ-algorithms for the bilinear problem (2.4) is called the λ-rank of the problem (2.4). (Originally the λ-rank has been introduced under the name "border rank", see [BCLR] and [B80].)

In particular, if (2.4) represents $m \times n \times p$ MM, that is, if $\alpha = (i,j)$, $\beta = (j,k)$, $\gamma = (k,i)$, $b_\gamma(X,Y) = \sum_{j=0}^{n-1} x_{ij} y_{jk}$, i,j,k range from 0 to m-1, n-1, p-1, respectively, then (6.1) turns into the general bilinear λ-algorithm of λ-rank M and of degree d for such a problem (m,n,p) over a ring F,

$$\sum_j x_{ij}y_{jk} = \lambda^{-d} \sum_{q=0}^{M-1} f''(k,i,q,\lambda)L_q L_q' + \lambda\, P \text{ for all } k,i. \tag{6.3}$$

Here

$$L_q = \sum_{i,j} f(i,j,q,\lambda)x_{ij},\ L_q' = \sum_{j,k} f'(j,k,q,\lambda)y_{jk},\ q=0,1,\ldots,M-1, \tag{6.4}$$

and P,f,f',f'' are defined as in the general case (6.1),(6.2).

The general definition also includes the cases of bilinear λ-algorithms for Partial MM and for Disjoint MM. In particular, the identities (4.5),(4.6) define a bilinear λ-algorithm of λ-rank mnp+mn+np and of degree 1 for two disjoint matrix products. In Example 4.1 in Section 4 we encountered a bilinear λ-algorithm of λ-rank 5 and degree 1 for Partial MM, which we immediately transformed into a bilinear λ-algorithm of λ-rank 10 and degree 1 for 2 X 2 X 3 MM.

The bilinear λ-algorithms can be considered as conventional bilinear algorithms over the ring of all finite power series in λ that include negative power of λ where the results of the computation are reduced modulo λ. Indeed, write modulo λ at the place of $+\lambda P$ in (6.1),(6.3), so that only λ-free terms are left on the right sides and the identities (6.1),(6.3) are rewritten as follows,

$$b_\gamma(X,Y) = \lambda^{-d} \sum_q f''(\gamma,q,\lambda)L_q L_q' \text{ modulo } \lambda \text{ for all } \gamma, \tag{6.5}$$

$$\sum_j x_{ij}y_{jk} = \lambda^{-d} \sum_q f''(k,i,q,\lambda)L_q L_q' \text{ modulo } \lambda \text{ for all } k,i. \tag{6.6}$$

Here f'', L_q, L_q' have the same meaning as in (6.1),(6.3). Recall Example 4.1 and the λ-algorithms (4.5),(4.6) and note that we do not assume that any of f'', L_q and L_q' have been reduced modulo λ but only that the right sides of (6.5) and (6.6) have been reduced modulo λ for each γ or for each pair k,i, respectively.

Another equivalent representation of bilinear λ-algorithms is obtained if we multiply (6.5) and (6.6) by λ^d and use the following resulting identities modulo λ^{d+1} at the places of (6.1),(6.5) and (6.3),(6.6), respectively.

$$\lambda^d\, b_\gamma(X,Y) = \sum_{q=0}^{M-1} f''(\gamma,q,\lambda)\, L_q L_q' \text{ modulo } \lambda^{d+1} \text{ for all } \gamma. \tag{6.7}$$

$$\lambda^d \sum_j x_{ij}y_{jk} = \sum_{q=0}^{M-1} f''(k,i,q,\lambda)\, L_q L_q' \text{ modulo } \lambda^{d+1} \text{ for all } k,i. \tag{6.8}$$

The reduction modulo λ in (6.5),(6.6) as well as the multiplication by λ^d and the reduction modulo λ^{d+1} in (6.7),(6.8) enable us to get rid of the term λ P in the right sides of (6.1),(6.3). In the case of the λ-algorithms (4.5),(4.6), this means that the terms $\lambda \sum_j (x_{ij}+u_{jk})v_{ki}$ indeed vanish due to the reductions modulo λ or λ^2. Similarly for the algorithm of Example 4.1.

Remark 6.1. Actually using the representation (6.7) we interpret bilinear λ-algorithms for the evaluation of a given set of bilinear forms $\{b_\gamma(X,Y)\}$ over a ring F of constants as conventional bilinear algorithms for the evaluation of the set $\{\lambda^d b_\gamma(X,Y)\}$ where the coefficients f,f',f'' are taken from the ring of polynomials in λ modulo λ^{d+1} with coefficients from the ring F.

If in (6.7),(6.8) the degree d is 0, then the λ-algorithms turn into conventional bilinear algorithms. If d>0, then we also can come back to bilinear algorithms by increasing the rank at most $(d+1)^2$ times. Here we could simply refer to Remark 6.1 and Proposition 5.2. However, we will present a proof independent of Section 5. Start with the λ-algorithm (6.7) or (6.8) and represent L_q and L_q' for all q as polynomials in λ modulo λ^{d+1}. Reduce each arithmetical operation over polynomials in λ modulo λ^{d+1} to several operations over their coefficients. Each such coefficient is actually a λ-free linear form in the x-variables or in the y-variables. Then all linear operations over polynomials (that is, their multiplication by constants, addition, or subtraction) will be reduced to the similar linear operations over the coefficients. Each bilinear step of the λ-algorithm will be reduced to the problem PM(d), that is, to at most $(d+1)^2$ muultiplications of the coefficients of those two polynomials and to some additions, compare Example 2.2. Not counting additions, this amounts to at most $(d+1)^2$ multiplications of λ-free linear forms in the x-variables by λ-free linear forms in the y-variables, that is, to at most $(d+1)^2$ bilinear steps, all of which are λ-free. Thus we obtain a conventional (λ-free) bilinear algorithm of rank at most $M(d+1)^2$ that evaluates all of the coefficients of the λ-polynomials $\sum_q f''(\gamma,q,\lambda)L_qL_q'$ modulo λ^{d+1} for all γ. In particular, the coefficients of λ^d-terms of those polynomials, which are equal to the desired values of $b_\gamma(X,Y)$, are evaluated for all γ. Summarizing we obtain a bilinear algorithm of rank at most $M(d+1)^2$ for the given problem of MM. (Similarly for Disjoint MM.) The rank can be reduced to (2d+1) M if the computation is over a field of at least 2d+1 distinct constants because in that case we may apply the interpolation in order to perform bilinear steps of λ-algorithms as in the algorithm of Example 2.2. We arrived at the following result.

Proposition 6.1. A bilinear λ-algorithm of degree d and of λ-rank M for a bilinear problem (in particular, for the problems of MM or Disjoint MM) can be transformed into a conventional bilinear algorithm of rank at most $M(d+1)^2$ for the same problem (or even of rank at most $(2d+1)M$ for the same problem if the computation is over a field of constants that contains at least $2d+1$ elements).

If we apply Proposition 6.1 and transform, say, the λ-algorithm of λ-rank 10 and degree 1 for (2,2,3), see Example 4.1, into a conventional bilinear algorithm, then the rank will increase $(d+1)^2=4$ times (or $2d+1=3$ times if we work over the fields of constants with at least 3 elements). Of course, this would be too big a sacrifice, which would make the resulting algorithm inefficient. Thus at first we will raise a recursive construction as we did proving Theorem 2.1. We will rely on the following extension of Proposition 2.2 (compare also Proposition 10.1 and Remark 10.2 in Section 10 for a proof of that extension as a special case of a more general fact).

<u>Proposition</u> <u>6.2</u>. Every pair of bilinear λ-algorithms B^* and B of λ-ranks M^* and M and of degrees d^* and d for the problems (m^*,n^*,p^*) and (m,n,p), respectively, define a bilinear λ-algorithm $B^* \otimes B$ of λ-rank at most M^*M and of degree at most d^*+d for the problem (m^*m,n^*n,p^*p).

The proof is similar to the proof of Proposition 2.2. However, a preliminary transformation of the identities (6.8) is required due to the factor λ^d on their right sides. Thus represent the λ-algorithm B^* using the identity (6.8), multiply its both sides by λ^d, and derive that

$$\lambda^{d+d^*} \sum_j x_{ij}y_{jk} = \sum_{q=0}^{M^*-1} f^{*''}(k,i,q,\lambda)(\lambda^d L_q^* L_q^{*'}) \text{ modulo } \lambda^{d+d^*+1}.$$

Substitute $m \times n$ matrices for the x-variables and $n \times p$ matrices for the y-variables. Then apply the λ-algorithm B to the evaluation of all of the $m \times n$ by $n \times p$ matrix products $\lambda^d L_q^* L_q^{*'}$ for all q, see (6.8). ∎

<u>Remark</u> <u>6.2</u>. It is tempting to prove Proposition 6.2 by merely applying Proposition 2.2 to the evaluation over the rings of either finite or formal power series in λ as that was done in [P79],[S79],[S81]. This, however, does not give a complete proof. We cannot merely reduce the factors L_q, and L_q' modulo λ because we need to preserve some of their terms of the form $f\lambda x$ or $f'\lambda y$, see (4.5), (4.6), or Example 4.1. On the other hand, if we consider f'', L_q, L_q' as nonreduced modulo λ finite power series in λ, then we should show that $B^* \otimes B$, the product of the λ-algorithms B^* and B, does not include any terms of degree less than d^*+d+1 generated by the products $\lambda P^* f(i,j,q,\lambda) L_q L_q'$ and $f^*(i,j,q,\lambda) L_q^* L_q^{*'} \lambda P$. The necessity to prove the latter fact in order to substantiate Proposition 6.2 and consequently all exponents of [CW],[P79],[S79],[S81] was first noticed in [P81] where that simple fact was proven using the equations (2.12). <u>This remark does not influence our diagram in</u>

Section 3, which reflects the dates of the announcements of the new exponent rather than the dates where their final rigorous proofs were presented, provided, surely, that those exponents (or even better ones, as this occurred with the exponents from the papers [P78] and [P79]) can be easily deduced from the algorithmic designs presented simultaneously with the announcements.

Now let B be a λ-algorithm of λ-rank M and of degree d for the problem (m,n,p). Then start with B and raise the recursive construction of its powers $B^{\otimes h} = B^{\otimes (h-1)} \otimes B$, h=2,3,..., relying on Proposition 6.2. Then $B^{\otimes h}$ is a bilinear λ-algorithm of λ-rank at most M^h and of degree at most dh for the problem (m^h,n^h,p^h), h=1,2,... Apply Proposition 6.1 and obtain a conventional (λ-free) bilinear algorithm of rank at most $(dh+1)^2M^h$ for the problem (m^h,n^h,p^h). Apply Theorem 2.1 and derive the following relation,
$\omega \leq 3 \log((dh+1)^2M^h)/\log(mnp)^h = 3 \log M/\log(mnp) + 6 \log(dh+1)/(h \log(mnp))$
for h=1,2,... Choose $h \to \infty$ and derive the next extension of Theorem 2.1, due to [B80] where that extension was obtained using a different idea (of interpolation).

Theorem 6.3. Given a bilinear λ-algorithm (6.3) of λ-rank M and of degree d for the problem of $m \times n \times p$ MM, then the inequality (2.6) holds.

Comparing Theorems 2.1 and 6.3 we conclude that a bilinear algorithm of rank M and a bilinear λ-algorithm of rank M and of arbitrary degree d for the same problem of MM of any specific size (m,n,p) define the same bound (2.6) on the exponent of MM. In Sections 7 and 8 we will see that a λ-algorithm of λ-rank M and a conventional algorithm of rank M for the same problem of Disjoint MM also define the same exponent.

Remark 6.3. The application of Theorem 6.3 to the λ-algorithm of λ-rank 10 for the problem (2,2,3), see Example 4.1, immediately yields the exponent 2.779... that defines the asymptotic complexity of the exact evaluation of matrix products.

Remark 6.4. The above reduction of λ-algorithms to conventional ones becomes efficient for MM only if the matrices involved are of enormously large sizes. This is due to the increase of the rank $(d+1)^2$ times (or 2d+1 times if the constants are from a field of cardinality at least 2d+1) required for that reduction, compare Remark 5.2. In Sections 23 (see Theorem 23.7) and 33 (see Examples 33.1 and 33.2) we will see that λ-algorithms are efficient already for MM of moderate size if they are applied as approximation algorithms for small $|\lambda|$.

Finally we will summarize and generalize the approach of this section as follows. Given a fast irregular algorithm for MM whose costly transition to a desired regular form would make the algorithm too slow. Then at first recursively apply the irregular algorithm in order to accumulate its accelerating power and only after that make the transition to the desired regular form. We will call such a general approach the accumulation of the accelerating power by recursion (hereafter referred to as AAPR). In the sequel we will encounter some further applications of the idea of AAPR.

Remark 6.5. The specific feature of our recursive construction of λ-algorithms, comparing with other applications of AAPR, is that we initially transform the algorithms by multiplying all terms of the associated identities by λ^d in order to be able to make the recursive substitution of the basis λ-algorithms (6.8).

7. Strassen's Conjecture. Its Extended and Exponential Versions.

In order to substantiate the bounds (4.8),(4.9), we will define a recursive construction based on the λ-algorithm (4.5),(4.6) for Disjoint MM. We could speculate that, since the λ-algorithm (4.5),(4.6) of λ-rank $M_{m,n,p,\lambda}$ computes both disjoint matrix products XY and UV, then probably at least one of the two products, XY or UV, can be computed by a bilinear λ-algorithm of λ-rank not greater than $M_{m,n,p,\lambda}/2$. If such a λ-algorithm existed, we could apply Theorem 6.3 and derive (4.8). Generalizing that idea about the free transition from Disjoint MM to MM we come to the following famous conjecture due to V. Strassen (we will state that conjecture for conventional bilinear algorithms as this was done in the original paper [St73]): given a bilinear algorithm of rank M over a field of constants for two arbitrary disjoint bilinear problems (not even necessarily for the problems of MM), then there exist two bilinear algorithms of ranks M_1 and M_2 for each of those two problems such that $M=M_1+M_2$. If Strassen's conjecture were true, we would apply it to the algorithm (4.1),(4.2). Then applying Theorem 2.1 we would easily deduce (4.4). If we could extend the conjecture to the case of λ-algorithms, then we would derive (4.8) from (4.5),(4.6) using Theorem 6.3, compare below.

It "remains" to prove the conjecture and its extension. However, at this point our progress stops. To see that the conjecture is not easy to prove, consider the particular case of the bilinear algorithm (4.1),(4.2) where m=n=p=10. Then the algorithm has rank 1300 and computes two disjoint 10 X 10 matrix products. Therefore the conjecture would imply that $\rho(10,10,10) \leq 650$ while it is still unknown even if there exists a bilinear algorithm of rank 700 for 10 X 10 MM.

Furthermore if we allowed the exclusion of the terms $\lambda\sum_j(x_{ij}+u_{jk})v_{ki}$ that vanish as λ vanishes, then the bilinear λ-algorithm (4.6),(4.7) would be a formal counterexample to extended Strassen's conjecture. Indeed, let m=p, n=1. Then the identities (4.5),(4.6) define a bilinear λ-algorithm of λ-rank m(m+2) that computes both mX1 by 1Xm and 1Xm by mXm disjoint matrix products XY and UV with arbitrary precision. It is not hard to show, however, that at least m^2 bilinear steps or, equivalently, at least m^2 linearly independent terms $L_q L_q'$ are required in each bilinear λ-algorithm over an arbitrary field F that computes one of the two given matrix products with arbitrary precision, see Theorem 36.1 in Section 36. Since $2m^2 > m(m+2)$ for $m > 2$, this disproves the extension of Strassen's conjecture to the class of bilinear λ-algorithms. We will state the latter result as follows using the interpretation of bilinear λ-algorithms given in Remark 6.1.

Theorem 7.1, compare Remark 13.1 in Section 13. For an arbitrary field F, Strassen's conjecture does not hold over the class of bilinear algorithms that use the constants from the ring of polynomials in λ modulo λ^{d+1} over F where $d>0$.

Remark 7.1. In the presented counterexample to the extended Strassen's conjecture (and actually in all known counterexamples to that conjecture) the bilinear algorithms are defined over the rings that contain at least one divisor of zero. (In the above example, λ is a zero divisor.) Over the rings containing no zero divisors (fields, integral domains, see [vdW]), Strassen's conjecture is still open.

Theorem 7.1 suggests that we should look for a reduction of Disjoint MM to MM without using Strassen's conjecture. Of course, we may ignore all computed matrix products but one (say, use only the subalgorithm (4.5) and compute only the matrix product XY). However, such a trivial reduction of Disjoint MM to MM may cost a substantial part of the accelerating power of the original algorithm. (For instance, the rank of the subalgorithm (4.5) for XY equals $mnp+np$, which is a too large value for $m \times n$ by $n \times p$ MM.) In order to utilize our algorithm (4.5),(4.6), we will use AAPR, that is, we will first apply the algorithm recursively to accumulate its accelerating power. After several recursive steps the relative cost of the trivial reduction to MM will become almost negligible and practically will not influence the exponent. To see how this works out, consider the algorithm (4.1),(4.2) in the case where $m=n=p$. Note that $M_{n,n,n} = n^3+3n^2$ is even and $Q = M_{n,n,n}/2$ is integer for all n. The key step will consist in devising a sequence of recursive algorithms of ranks $2Q^h$ for two disjoint problems of $n^h \times n^h$MM, $h=2,3,\ldots$, see the mappings (8.6) in the next section. For each h apply such an algorithm to one of those two problems. Choose $h \to \infty$, apply Theorem 2.1, and derive that

$$\omega \leq \log Q/\log n, \quad Q = M_{n,n,n}/2 \text{ for all } n. \tag{7.1}$$

(7.1) amounts to (4.4) in the case $m=n=p$.

We will deduce (4.4) from (4.1),(4.2) for all m,n,p and (4.8) from (4.5),(4.6) for all m,n,p using AAPR along this line. On the other hand, if Strassen's conjecture held true, then the identities (4.1),(4.2), and the fact that the ranks are always integers would imply that

$$\text{minimum } \{ \rho(m,n,p), \rho(n,p,m)\} \le \lfloor M_{m,n,p}/2 \rfloor .$$

(Here and hereafter $\lfloor u \rfloor$ designates the greatest integer that does not exceed a real u, compare Notation 18.1.) Then (4.4) would follow by the virtue of Theorem 2.1.

Similarly if extended Strassen's conjecture had held, then (4.5),(4.6) would have implied that the λ-rank of at least one of the two problems (m,n,p) or (n,p,m) was bounded by $\lfloor M_{m,n,p,\lambda}/2 \rfloor$. Then (4.8) would follow by the virtue of Theorem 6.3. Actually (extended) Strassen's conjecture, if true, would lead to even stronger upper bounds whenever the values of $M_{m,n,p}$ (and/or $M_{m,n,p,\lambda}$, respectively) are odd because $\lfloor M_{m,n,p}/2 \rfloor = (M_{m,n,p}-1)/2$, $\lfloor M_{m,n,p,\lambda}/2 \rfloor = (M_{m,n,p,\lambda}-1)/2$ for odd $M_{m,n,p}$, $M_{m,n,p,\lambda}$.

If we "forgot" that the ranks and λ-ranks must be integers, then in the considered cases Strassen's conjecture (in its original and extended versions) would always lead to the same exponents as AAPR does. This observation can be generalized as an informal rule for bounding the exponent of MM whenever a bilinear algorithm or a bilinear λ-algorithm for the evaluation of two or several disjoint matrix products is given. The formal (but equivalent!) version of that rule takes the form of the following theorem, which will be called the <u>weak</u> or <u>exponential version of Strassen's conjecture</u>.

<u>Theorem 7.2</u>. ([S81]). Given a bilinear λ-algorithm of λ-rank M or a bilinear algorithm of rank M for S disjoint problems of MM, (m(s),n(s),p(s)), s=0,1,...,S-1, where M>S, then

$$\omega \le 3\tau, \quad \sum_{s=0}^{S-1} (m(s)n(s)p(s))^{\tau} = M. \tag{7.2}$$

<u>Remark 7.2</u>. In the sequel we will show, see Theorem 36.1 in Section 36, that the inequality M>S always holds in the above unless m(s)n(s)p(s)=1 for all s.

Let us show that Theorem 7.2 could be easily proven if we could use Strassen's conjecture and its extension. At first proceed independently of the conjecture. Just by using Theorem 2.1 derive that

$$\omega \le 3\tau(s), \ (m(s)n(s)p(s))^{\tau(s)} = \rho(m(s),n(s),p(s)), \ s=0,1,..., S-1.$$

Therefore

$$\sum_s (m(s)n(s)p(s))^{\omega/3} \leq \sum_s \rho(m(s),n(s),p(s)). \qquad (7.3)$$

If at this point the conjecture is applied to the given algorithm, then it follows that

$$\sum_s \rho(m(s),n(s),p(s)) \leq M. \qquad (7.4)$$

(7.2) immediately follows from (7.3) and (7.4). Similarly if we start with the given λ-algorithm and apply the extended conjecture. ∎

In the next two sections we will prove Theorem 7.2 by applying AAPR and we will not refer to the conjecture at all. In our proof we will follow [P81] with the key step similar to one described above for deriving the bound (4.4) from the algorithm (4.1),(4.2). However, a substantial simplification of our proof below against one from [P81] is due to the formidable but actually very helpful notation that we will borrow from [S81] and will introduce in the beginning of the next section. In order to make the ideas more transparent, we will start with the two simpler particular cases where the basis algorithms are defined by (4.1),(4.2) and (4.5),(4.6). As is usual with the asymptotically fast recursive algorithms for MM defined via AAPR, the resulting acceleration of MM will have no practical value because it starts being efficient only where the matrices to be multiplied become very large, compare Remarks 5.2 and 6.4.

Those readers who are interested in the designs of fast algorithms rather than in their justification, may go directly to Section 10.

8. Recursive Algorithms for MM and for Disjoint MM (Definitions, Notation, and Two Basic Facts).

In order to substantiate the sketchy argument suggested in the previous section for proving the bound (4.4) (compare (7.1)), and to extend it to the proof of Theorem 7.2, we need to define recursive algorithms for Disjoint MM. In this section we will do that by implicitly using (tensor) products of two or several problems of Disjoint MM and by extending that concept in order to define the products of bilinear algorithms for such problems, compare Remark 2.4. Our main objective is to derive Proposition 8.1 below, which is an extension of Proposition 2.2' to the case of Disjoint MM. (A different derivation of all three Propositions 2.2',6.2, and 8.1 will be given in Section 10, see Remark 10.2.) Those readers who feel themselves well-prepared may try to deduce such an extension themselves. (They also might find it interesting to examine Definition 8.1 below.) We will proceed slowly but systematically with illustrations and demonstrations of our approach. At first we will introduce the notation promised at the end of Section 7 and taken from [S81].

Let hereafter $t = \bigoplus_s (m(s),n(s),p(s))$ designate the problem of the evaluation of the set of S matrix products X(s)Y(s) where X(s) and Y(s) are disjoint m(s) X n(s) and n(s) X p(s) matrices, respectively, that is, their entries are indeterminates, and where s ranges from 0 to S-1. This is the general problem of Disjoint MM. In particular, if m(s)=m, n(s)=n, p(s)=p are invariant in s, then we will write that $t = S \odot (m,n,p)$, compare (2.8)-(2.10). For S=1 we come back to the problem $(m,n,p) = 1 \odot (m,n,p)$ of mXnXp MM.

Our definitions of bilinear algorithms (Definition 2.2) and of bilinear λ-algorithms (Definition 6.1) can be applied to the particular case of Disjoint MM, which in turn generalizes the case of MM (S=1), as we could have just seen. The algorithm (4.1),(4.2) and the λ-algorithm (4.5),(4.6) are two particular instances of algorithms and λ-algorithms for Disjoint MM where S=2, X(0)=X, Y(0)=Y, X(1)=U, Y(1)=V, m(0)=m, n(0)=n, p(0)=p, m(1)=n, n(1)=p, p(1)=m and where M=mnp+mn+np+pm and M=mnp+mn+np, d=1, respectively. We will designate a bilinear algorithm of rank M for the problem $\bigoplus_S (m(s),n(s),p(s))$ of Disjoint MM by the mapping

$$\bigoplus_s (m(s),n(s),p(s)) \leftarrow M \odot (1,1,1) \tag{8.1}$$

and a bilinear λ-algorithm of rank M and degree d for the same problem by the mapping

$$\lambda^d \odot (\bigoplus_s (m(s),n(s),p(s))) \leftarrow M \odot (1,1,1). \tag{8.2}$$

Then again (compare our comments to the mapping (2.8)) the notation (8.1),(8.2) indicates that it suffices to perform the M main (bilinear) steps, that is, the M multiplications of L_q by L_q' (each considered as an 1X1 MM) in order to evaluate the S matrix products X(s)Y(s) or $\lambda^d X(s)Y(s)$, provided that the linear steps (of the evaluation of all linear combinations of the x-variables, of the y-variables, and of the products $L_q L_q'$) are considered free (the latter assumption can be motivated by Theorems 2.1 and 6.3). In particular the algorithms (4.1),(4.2) and (4.5),(4.6) are designated

$$(m,n,p) \oplus (n,p,m) \leftarrow (mnp+mn+np+pm) \odot (1,1,1)$$

and

$$\lambda \odot (m,n,p) \oplus (n,p,m)) \leftarrow (mnp+mn+np) \odot (1,1,1),$$

respectively, while the bilinear algorithm (2.2),(2.3), in the case where $m=n=p$, is represented by the following mapping,

$$(n,n,n) \leftarrow M \odot (1,1,1). \tag{8.3}$$

For the sake of completeness, we will present the following generalization of Definitions 2.2 and 6.1. (We will not actually use that generalization in our study of MM.)

Definition 8.1. Any set of bilinear forms in the x-variables and in the y-variables is called a bilinear problem. (Examples 2.1 and 2.2, MM, Disjoint MM and Partial MM, see Example 4.1, represent some particular instances of bilinear problems.) Let a combination of three linear transformations (i) of the x-variables, (ii) of the y-variables, and (iii) of the bilinear forms of a given bilinear problem t' define a new bilinear problem t. Then such a combination of the three linear transformations is called a bilinear algorithm that reduces t to t' or, equivalently, a bilinear mapping of t' into t designated by $t \leftarrow t'$. In the case where the coefficients of the given bilinear form are polynomials in λ, d is a nonnegative integer, and transformations (i), (ii), (iii) are always followed by the reduction of the given bilinear forms modulo λ^{d+1}, the combination of such four transformations (that is, of the three linear transformations of variables and of the reduction of the given bilinear problem modulo λ^{d+1}) will be called a bilinear λ-algorithm (or λ-mapping) of degree d that reduces t to t' and maps t' into t.

Next we will apply the notation (8.1),(8.2) in order to define the recursive construction of bilinear algorithms and of bilinear λ-algorithms for Disjoint MM. Substitute nXn matrices for the input-variables and change the sizes of the matrices on the left and on the right sides of the mapping (8.1), so that (8.1) takes the following form,

$$\bigoplus_{s} (m(s)n,n(s)n,p(s)n) \leftarrow M \odot (n,n,n).$$

This mapping will be considered as the product of the algorithm (8.1) and of the following trivial algorithm (mapping),

$$(n,n,n) \leftarrow (n,n,n), \tag{8.4}$$

compare Remark 2.4 in Section 2. The algorithm (8.4) can be interpreted as the <u>trivial reduction</u> of the problem of nXn MM to itself.

For illustration, we will first consider the particular case where we start with the algorithm (8.3) for MM. We will follow the proof of Theorem 2.1 in the case where m=n=p but we will simplify the matter by not caring about the cost-free linear operations. In this way we will multiply the mapping (8.3) by the trivial reduction mapping (8.4) and derive the algorithm that operates over nXn matrices.

$$(n^2,n^2,n^2) \leftarrow M \odot (n,n,n).$$

Perform each bilinear step of the latter algorithm using the algorithm (8.3) and obtain the bilinear algorithm

$$(n^2,n^2,n^2) \leftarrow M^2 \odot (1,1,1),$$

which can be considered the square of (8.3).

Similarly define all powers of the algorithm (8.3),

$$(n^h,n^h,n^h) \leftarrow M^h \odot (1,1,1),\ h=1,2,3,\ldots$$

Next define similar construction starting with an algorithm for Disjoint MM. Then again substitute matrices of appropriate sizes for the variables. For instance, if you start with the algorithm (8.1) in the case where m=n=p,

$$2 \odot (n,n,n) \leftarrow M_{n,n,n} \odot (1,1,1), \tag{8.5}$$

then substitute nXn matrices for all variables and consequently obtain the algorithm

$$2 \odot (n^2,n^2,n^2) \leftarrow M_{n,n,n} \odot (n,n,n).$$

Note that the values $Q=M_{n,n,n}/2=(n^3+3n^2)/2$ are integers for all n, apply the former algorithm in order to perform <u>each of the</u> Q <u>pairs of bilinear steps</u> of the latter algorithm, and derive that

$$2 \odot (n^2,n^2,n^2) \leftarrow Q \odot (2 \odot (n,n,n)) \leftarrow Q\, M_{n,n,n} \odot (1,1,1).$$

This way, starting with the algorithm (8.5), recursively define the sequence of bilinear algorithms

$$2 \odot (n^h,n^h,n^h) \leftarrow 2\, Q^h \odot (1,1,1),\ h=1,2,\ldots \qquad (8.6)$$

(Then the bound (7.1) will immediately follow as this was explained in Section 7. Indeed, reduce (8.6) to the algorithm $(n^h,n^h,n^h) \leftarrow 2Q^h \odot (1,1,1)$, $h=1,2,\ldots$, apply Theorem 2.1, and let $h \to \infty$.)

The above demonstrations show how to define recursive algorithms starting with a given algorithm for Disjoint MM. We will formally reduce such recursive constructions to the recursive applications of two general Propositions 8.1 and 8.2 below. In order to come to Proposition 8.1, which extends Proposition 2.2' to the case of Disjoint MM, start with the general algorithm (8.1) for Disjoint MM and substitute rectangular matrices of the sizes $m(r)Xn(r)$ for the x-variables and of the sizes $n(r)Xp(r)$ for the y-variables of that algorithm. Repeat such substitutions for all S values of r choosing $r=0,1,\ldots,S-1$. Then the S resulting algorithms (mappings)

$$\bigoplus_s (m(r)m(s),n(r)n(s),p(r)p(s)) \leftarrow M \odot (m(r),n(r),p(r)),\ r=0,1,\ldots,S-1,$$

can be considered as the S products of the algorithm (mapping) (8.1) and of each of the S trivial mappings $(m(r),n(r),p(r)) \leftarrow (m(r),n(r),p(r))$ for $r=0,1,\ldots,S-1$. Combine the latter S products into the new algorithm

$$\bigoplus_{r,s} (m(r)m(s),n(r)n(s),p(r)p(s)) \leftarrow M \odot \Big(\bigoplus_{r=0}^{S-1} (m(r),n(r),p(r))\Big). \qquad (8.7)$$

Here and hereafter in this section we let r and s range from 0 to S-1. Thus the solution of the problem of Disjoint MM on the left side has been reduced to the solution of the M problems on the right side. (The mapping (8.7) can be considered the product of (8.1) and of the trivial mapping $\bigoplus_r (m(r),n(r),p(r)) \leftarrow \bigoplus_r (m(r),n(r),p(r))$. Some readers might be motivated to verify the laws of arithmetic for the products $\otimes$ and $\odot$ and for the sums $\oplus$.) It remains to apply the original algorithm (8.1) to all of those M problems on the right side of

(8.7) and derive the algorithm

$$\bigoplus_{r,s} (m(r)m(s),n(r)n(s),p(r)p(s)) \leftarrow M^2 \odot (1,1,1),$$

which can be considered the square of the algorithm (8.1). Similarly, if we are given the algorithm (8.1) and the algorithm

$$\bigoplus_{s^*} (m^*(s^*),n^*(s^*),p^*(s^*)) \leftarrow M^* \odot (1,1,1),$$

where s^* ranges from 0 to S^*-1, then we can derive the following algorithm, which we consider the product of the latter algorithm and of the algorithm (8.1),

$$\bigoplus_{s^*,s} (m^*(s^*)m(s),n^*(s^*)n(s),p^*(s^*)p(s)) \leftarrow M^* M \odot (1,1,1).$$

Summarizing we obtain the following generalization of Proposition 2.2', compare also Remark 10.1 in Section 10.

Proposition 8.1. *Any two bilinear mappings (algorithms)* $t^* \leftarrow t^{*'}$ *and* $t \leftarrow t'$ *define their product*, that is, the bilinear mapping (algorithm)

$$t^* \otimes t \leftarrow t^{*'} \otimes t',$$

where the *product of two problems of MM or of Disjoint MM* is defined by the following formulae.

$$(m^*,n^*,p^*) \otimes (m,n,p) = (m^*m,n^*n,p^*p), \tag{8.8}$$

$$\{\bigoplus_{s^*,s} (m^*(s^*),n^*(s^*),p^*(s^*))\} \otimes \{\bigoplus_{s}(m(s),n(s),p(s))\} =$$

$$\bigoplus_{s^*,s} (m^*(s^*)m(s),n^*(s^*)n(s),p^*(s^*),p(s)). \tag{8.9}$$

Similarly any two given bilinear λ-algorithms (mappings) $\lambda^{d^*} \odot t^* \leftarrow t^{*'}$ and $\lambda^d \odot t \leftarrow t'$ define their product, that is, the bilinear λ-algorithm (mapping)

$$\lambda^{d^*+d} \odot t^* \otimes t \leftarrow t^{*'} \otimes t'.$$

Actually we have formally derived Proposition 8.1 only for bilinear algorithms. However, the case of bilinear λ-algorithms can be handled similarly. Indeed, it suffices to work with the λ-algorithms as with conventional bilinear algorithms over the ring of polynomials in λ modulo λ^g for appropriate g (see Remark 6.1), using preliminary multiplications by the appropriate powers of λ before each substitution of such a λ-algorithm (compare the proof of Proposition 6.2). For instance, should the λ-algorithm (6.8) be substituted into the right side of (8.7), then as a preliminary multiply both sides of (8.7) by λ^d.■

Remark 8.1. Of course, the definition of the products is immediately extended to the powers $t^{\otimes 1} = t$, $t^{\otimes(h+1)} = t^{\otimes h}\otimes t$, h=1,2,... As follows from (8.8),(8.9), each of the three coordinates of the power of the problem $t = \bigoplus_s (m(s),n(s),p(s))$ can be expanded using the familiar formulae for the expansion of the powers of the linear form $\sum_s m(s)x_s$, $\sum_s n(s)y_s$, $\sum_s p(s)z_s$. In particular, we have that

$$(S \odot (m,n,p))^{\otimes h} = S^h \odot (m^h,n^h,p^h), \tag{8.10}$$

$$((m,n,p) \oplus (n,p,m))^{\otimes h} = \bigoplus_{g=0}^{h} C(h,g)\ (m^g n^{h-g}, n^g p^{h-g}, p^g m^{h-g}), \tag{8.11}$$

where C(h,g) are the binomial coefficients h!/(g!(h-g)!), g=0,1,...,h; h=1,2,...

We have derived Proposition 8.1 for the classes of bilinear algorithms and of bilinear λ-algorithms for MM and Disjoint MM and also for the trivial reduction algorithms of the form $(m,n,p) \leftarrow (n,n,p)$. These are the two cases that we will need to consider in our recursive constructions. At the final step of AAPR (see the previous and the next Sections 7 and 9) we will use the trivial reduction algorithms (mappings) of the forms

$$t \leftarrow t \oplus t^*,\quad \lambda^d \odot t \leftarrow \lambda^d \odot (t \oplus t^*) \tag{8.12}$$

where t and t^* may represent the problems of MM and/or Disjoint MM. The mappings (8.12) mean that if t and t^* are solved, then t is solved.

Remark 8.2. The readers may confine themselves to considering only the mappings of just listed kinds that represent MM, Disjoint MM, or trivial reductions (8.12). On the other hand, Proposition 8.1 (and consequently our whole study of recursive algorithms) can be immediately generalized to the case where $t^*, t, t^{*\prime}, t'$ represent four arbitrary bilinear problems where the meaning of the mapping is given by Definition 8.1 and where the two products $t^* \otimes t$ and $t^{*\prime} \otimes t'$ represent the two bilinear problems associated with the two products of the two pairs of tensors,

(t^*,t) and $(t^{*'},t')$. Those readers who do not mind to deal with tensors and their products will probably prefer to have such a unified point of view at the recursive bilinear algorithms and bilinear λ-algorithms for all bilinear problems, compare Remarks 2.4 in Section 2 and 10.2 in Section 10.

For the bilinear algorithms and bilinear λ-algorithms that we have defined (including ones defined in Definition 8.1) we obviously have the following rule of transitivity of bilinear algorithms.

Proposition 8.2. Every two bilinear algorithms or bilinear λ-algorithms of the form $t \leftarrow t'$ and $t' \leftarrow t''$ define the new bilinear algorithm or bilinear λ-algorithm $t \leftarrow t''$ (called their composition).

Proposition 8.1 and 8.2 enable us to represent the construction of recursive algorithms as a sequence of multiplications and compositions of mappings. (To see an illustration, recall how we derived recursive algorithms from the algorithms (8.3) and (8.5) and also compare the recursive algorithms in the next section.)

9. Some Applications of the Recursive Construction of Bilinear Algorithms.

In this section we will apply our definition of the recursive construction via the multiplications and the compositions of mappings (see Propositions 8.1 and 8.2) in order to substantiate the bounds (4.8),(4.9) and to prove Theorem 7.2. (In our proof we will follow the line of [P81].)

At first we will illustrate our approach by revisiting the reduction of the proof of Theorem 2.1 for arbitrary m,n,p to the case m=n=p. Again we will use the two dual mappings, that is, the two algorithms of rank M,

$$(n,p,m) \leftarrow M \odot (1,1,1), \quad (p,m,n) \leftarrow M \odot (1,1,1)$$

whenever we are given the mapping $(m,n,p) \leftarrow M \odot (1,1,1)$, see Theorem 2.3. Multiplying all three of those mappings (compare Proposition 2.2') we obtain the following algorithm for a square MM,

$$(mnp,npm,pmn) \leftarrow M^3 \odot (1,1,1). \quad \blacksquare$$

Next we will prove the basic result that we will need for AAPR whenever we start with an algorithm for Disjoint MM.

Proposition 9.1. Let S divide M, so that Q=M/S is an integer. Then for all natural d the bilinear mapping (which is a λ-algorithm for Disjoint MM)

$$S \lambda^d \circledcirc (m,n,p) \leftarrow M \circledcirc (1,1,1), \ mnp > 1 \qquad (9.1)$$

defines the following bound,

$$\omega \leq 3 \log (M/S)/\log(mnp). \qquad (9.2)$$

Remark 9.1. Proposition 9.1, as well as other results of this section, can be applied in both cases where we deal with bilinear algorithms (d=0) and bilinear λ-algorithms (d>0).

Our proof of Proposition 9.1 will illustrate the idea of AAPR and the application of the products and the compositions of bilinear mappings. We will start with the algorithm (9.1) for Disjoint MM, perform several recursive steps, and finally reduce the resulting algorithm for Disjoint MM to an algorithm for MM. We will proceed similarly to the derivation of the algorithm (mapping) (8.6), in the previous section. For simplicity, we will assume that d=0. This will be no actual restriction because we will rely only on Theorem 6.3 and Propositions 8.1,8.2, which can be equally applied to both cases d>0 (λ-algorithms) and d=0 (conventional algorithms).

At first multiply (9.1) by the trivial mapping $(m,n,p) \leftarrow (m,n,p)$, then compose the resulting mapping with (9.1) (or, informally speaking, substitute (9.1) on the right side), and derive the following sequence of mappings,

$$S \circledcirc (m^2,n^2,p^2) \leftarrow Q\, S \circledcirc (m,n,p) = Q \circledcirc (S \circledcirc (m,n,p))$$

$$\leftarrow Q\, M \circledcirc (1,1,1) = Q^2 \circledcirc (S \circledcirc (1,1,1)).$$

Again multiply that sequence by the same trivial mapping, substitute (9.1) on the right side, and so on. After h-1 steps obtain the following composition of mappings (which is still an algorithm for Disjoint MM but not for MM).

$$S \circledcirc (m^h,n^h,p^h) \leftarrow Q^h\, S \circledcirc (1,1,1).$$

Compose the latter mapping with the trivial reduction mapping

$$(m^h,n^h,p^h) \leftarrow S \circledcirc (m^h,n^h,p^h).$$

Obtain the resulting mapping (which is already an algorithm for MM)

$$(m^h,n^h,p^h) \leftarrow Q^h S \odot (1,1,1)$$

for arbitrary h.

Apply Theorem 2.1 (or Theorem 6.3 if you deal with λ-algorithms) and derive that

$$\omega \leq \omega(h) = 3 \log(Q^h S)/\log(mnp)^h =$$

$$3 \log Q/\log(mnp)+3 \log S/(h \log(mnp)) \text{ for all } h.$$

Hence

$$\omega \leq \lim_{h\to\infty} \omega(h) = 3 \log Q/\log(mnp).$$

This proves Proposition 9.1 because $Q=M/S$. ∎

Remark 9.2. We have just proven that $c(h)N^{\omega(h)}$ arithmetical operations suffice for NXN MM. Here $c(h)$ is a constant for a fixed h but $c(h)$ is unbounded for $h\to\infty$. Therefore $O(N^{\omega+\varepsilon})$ arithmetical operations suffice for NXN MM for arbitrary positive ε but not necessarily for $\varepsilon=0$. Thus (9.2) defines a truly limiting exponent. We will see that the same is true for all exponents derived from bilinear λ-algorithms and from conventional bilinear algorithms for Disjoint MM unless we deal just with conventional algorithms for MM. Moreover, it is possible to strengthen that observation unless the exponent of MM is equal to 2 (see Theorems 15.1 and 15.2 in Section 15).

To prove the bound (4.8), we will reduce it to Proposition 9.1. The next result will be our first step in that direction. Here again we will use the idea of AAPR and the recursive construction in the form of the products (in this case of the powers) and of the compositions (substitutions) of bilinear mappings. We will proceed similarly also at the final step of the proof of (4.8).

Proposition 9.2. (9.1) implies (9.2) even if S does not divide M.

Proof. Recursively apply Proposition 8.1, recall (8.10), and derive from (9.1) that

$$\lambda^{dh} \odot S^h \odot (m^h,n^h,p^h) \leftarrow M^h \odot (1,1,1),\ h=1,2,\ldots$$

Then trivially reduce the left sides and obtain that

$$\lambda^{dh} \otimes M^g \otimes (m^h, n^h, p^h) \leftarrow M^h \otimes (1,1,1)$$

where

$$M^g \le S^h < M^{g+1}, g \le h.$$

Apply Proposition 9.1 to the latter mappings, assume that $h \to \infty$, and obtain that

$$\omega \le \omega(h) = 3 \log(M^h/M^g)/\log(mnp)^h = 3 \log(M^h/S^h)/\log(mnp)^h + O(1/h) =$$

$$3 \log(M/S)/\log(mnp) + O(1/h).$$

Let $h \to \infty$ and deduce (9.2). ■

To derive (4.8) in the general case, that is, for all m,n,p, we need one more application of the idea of AAPR.

Recursively apply the algorithm (4.5),(4.6) and obtain its powers. This defines the following sequence of the associated mappings,

$$\lambda^h \otimes ((m,n,p) \oplus (n,p,m))^{\otimes h} \leftarrow (M_{m,n,p,\lambda})^h \otimes (1,1,1), \ h=1,2,\ldots$$

Expand the left sides (see (8.11)), apply trivial reductions, and deduce the following mappings for g=0,1,..,h and h=1,2,...,

$$\lambda^h \ C(h,g) \otimes (m^g n^{h-g}, n^g p^{h-g}, p^g m^{h-g}) \leftarrow (M_{m,n,p,\lambda})^h \otimes (1,1,1).$$

By the virtue of Proposition 9.2, it follows that

$$\omega \le 2 \log(M_{m,n,p,\lambda})^h/C(h,g))/\log(mnp)^h = 3 \log(M_{m,n,p,\lambda}/C(h,g)^{1/h})/\log(mnp).$$

Choose h=2g. Then $\lim_{h \to \infty} C(h,g)^{1/h} = 2.$

Substitute this relation into the latter mapping (where let $g=h/2$, $h \to \infty$) and deduce the bound (4.8) and consequently the bound (4.9) for arbitrary m,n,p. ■

We have derived the bound (4.8) starting with the basic λ-algorithm (4.5),(4.6). If we start with the general algorithm (8.1) for Disjoint MM, we can similarly prove (7.2) without using Strassen's conjecture. Here is the <u>sketch of the proof</u>. Expand the h-th tensor power of the algorithm (8.1), then choose an appropriate trivial reduction to the mapping of the form (9.1), apply Proposition 9.2, and finally choose $h \to \infty$. Let us supply some details.

<u>Proof of Theorem 7.2</u>. At first recursively apply Proposition 8.1 to the mapping (8.1) and derive the mappings

$$\lambda^{dh} \odot t^{\otimes h} \leftarrow M^h \odot (1,1,1),\ h=1,2,\dots \tag{9.3}$$

where $t = \bigoplus_s (m(s),n(s),p(s))$. Here and hereafter in this section we will keep assuming that s ranges from 0 to S-1.

Expand $t^{\otimes h}$ and obtain that $t^{\otimes h} = \bigoplus_{\underline{r} \in \underline{R}} Q(\underline{r}) \odot t(\underline{r})$. Here $\underline{R} = \underline{R}_{h,S}$ is the set of all S-dimensional vectors $\underline{r} = [r(0),r(1),\dots,r(S-1)]$ such that all r(s), $s=0,1,\dots,S-1$, are nonnegative integers and $\sum_{s=0}^{S-1} r(s) = h$;

$$\left.\begin{aligned} t(\underline{r}) &= \bigotimes_s (m(s),n(s),p(s))^{\otimes r(s)} = \Big(\prod_s m(s)^{r(s)}, \prod_s n(s)^{r(s)}, \prod_s p(s)^{r(s)}\Big); \\ Q(\underline{r}) &= Q_h(\underline{r}) = h!/\prod_s r(s)!, \quad \sum_{\underline{r} \in \underline{R}} Q(\underline{r}) = S^h. \end{aligned}\right\} \tag{9.4}$$

As follows from the definition of $\underline{R}$,

$$|\underline{R}| \le (h+1)^{S-1} \tag{9.5}$$

where $|\underline{R}|$ designates the cardinality of the set $\underline{R}$.

Next combine (9.3) with the trivial reduction $Q(\underline{r}) \odot t(\underline{r}) \leftarrow t^{\otimes h}$ for an arbitrary vector $\underline{r} \in \underline{R}$, apply Propositions 8.2, and derive that

$$\lambda^{dh} \odot Q(\underline{r}) \odot t(\underline{r}) \leftarrow M^h \odot (1,1,1),\ h=1,2,\dots$$

Application of Proposition 8.2 yields the bound

$$\left.\begin{aligned} &\omega \leq 3\tau_h(\underline{r}), \quad \tau_h(\underline{r}) = \log(M^h/Q(\underline{r}))/\log G(\underline{r}), \\ &G(\underline{r}) = \prod_s (m(s)n(s)p(s))^{r(s)}, \quad h=1,2,\ldots, \text{ for all } \underline{r} \in \underline{R}. \end{aligned}\right\} \quad (9.6)$$

Let $\underline{r} = \underline{r}^* = [r^*(0), r^*(1), \ldots, r^*(S-1)]$ be such that

$$Q(\underline{r}^*)(G(\underline{r}^*))^{\tau} \geq Q(\underline{r})(G(\underline{r}))^{\tau} \text{ for all } \underline{r} \in \underline{R}. \quad (9.7)$$

Here and hereafter τ designates the positive solution to the equation (7.2) (which is obviously unique). Then Theorem 7.2 will immediately follow from (9.6) and from the next auxiliary result.

Proposition 9.3. $\lim_{h \to \infty} \tau_h(\underline{r}^*) \leq \tau$ if $M > S$, if $\tau_h(\underline{r})$ is defined by (9.6), and if $\underline{r}^*$ is defined by (9.7).

Proof of Proposition 9.3. Equate the h-th powers of the left and right sides of the equation (7.2) and obtain that

$$(\sum_s (m(s)n(s)p(s))^{\tau})^h = M^h \text{ for all } h. \quad (9.8)$$

Expand the left side of (9.8), substitute the values $Q(\underline{r})$ and $G(\underline{r})$ defined by (9.4),(9.6), and derive that

$$\sum_{\underline{r} \in \underline{R}} Q(\underline{r})(G(\underline{r}))^{\tau} = M^h \text{ for all } h.$$

Combine the latter equation with the inequalities (9.5) and (9.7) and deduce that

$$Q(\underline{r}^*)(G(\underline{r}^*))^{\tau} \geq M^h/|\underline{R}| \geq M^h/(h+1)^{S-1},$$

Therefore

$$(G(\underline{r}^*))^{\tau} \geq (M^h/Q(\underline{r}^*))/(h+1)^{S-1}, \quad (9.9)$$

$$\tau \log G(\underline{r}^*) \geq \log(M^h/Q(\underline{r}^*)) - (S-1)\log(h+1).$$

Consequently (see (9.6))

$$\tau \geq \tau_h(\underline{r}^*) - (S-1) \log (h+1)/\log G(\underline{r}^*). \tag{9.10}$$

As follows from (9.4), $Q(\underline{r}^*) \leq S^h$. Combine the latter inequality with (9.9) and derive that

$$G(\underline{r}^*) \geq (M/S)^{h/\tau} / (h+1)^{(S-1)/\tau}.$$

Therefore $G(\underline{r}^*)$ grows exponentially in h as $h \to \infty$ since M>S. Hence $\lim_{h \to \infty} ((S-1) \log (h+1)/\log G(\underline{r}^*)) = 0$. This implies Proposition 9.3, see (9.10). ∎

10. Trilinear Versions of Bilinear Algorithms and of Bilinear λ-algorithms. Duality. Recursive Trilinear Algorithms.

So far we thoroughly exploited the power of our basic algorithm (4.1),(4.2) and of its λ-improvement (4.5),(4.6) using the idea of AAPR and developing appropriate techniques. However, the basic algorithm itself has appeared ad hoc in Section 4. Here and in the next section we will describe a general approach to designing efficient basis algorithms. As the first step, we will define the equivalent trilinear version of bilinear algorithms and of λ-algorithms in the form of some special decompositions of trilinear forms. For illustration, we will start with the trilinear version of the rank 3 algorithm for the complex products (see our Example 2.1 in Section 2). We will introduce two auxiliary variables z_0, z_1, multiply the real part identity $x_0y_0 - x_1y_1 = (x_0-x_1)(y_0+y_1) - x_0y_1 + x_1y_0$ by z_0, and the imaginary part identity $x_0y_1 + x_1y_0 = x_0y_1 + x_1y_0$ by z_1, and sum those two products. The resulting trilinear identity

$$x_0y_0z_0 - x_1y_1z_0 + x_0y_1z_1 + x_1y_0z_1 = \tag{10.1}$$

$$(x_0 - x_1)(y_0 + y_1)z_0 + x_0y_1(-z_0 + z_1) + x_1y_0(z_0 + z_1)$$

represents the original algorithm of rank 3 for the complex product. In order to go back from (10.1) to the bilinear identities, just equate the coefficients of each of z_0 and z_1 on both sides of (10.1).

This can be immediately extended to the trilinear versions of all bilinear algorithms and of all bilinear λ-algorithms for all bilinear problems. In particular here is the trilinear version of the bilinear algorithm (2.2),(2.3) for MM.

$$\mathrm{Tr}(XYZ) = \sum_{i,j,k} x_{ij}y_{jk}z_{ki} = \sum_{q=0}^{M-1} L_qL_q'L_q'', \quad L_q'' = \sum_{k,i} f''(k,i,q)z_{ki}. \tag{10.2}$$

Here L_q and L_q' for all q are defined by (2.2); TrU is the trace of a matrix $U=[u_{qs}]$, $\mathrm{Tr}U = \sum_s u_{ss}$; $Z = [z_{ki}]$ is the matrix of the auxiliary variables z_{ki}. (We prefer to write z_{ki} rather than z_{ik}.) The identity (10.2) can be obtained as the sum of the identities (2.2) multiplied by z_{ki} for all k,i. For instance, Strassen's algorithm for 2X2 MM can be rewritten as the following decomposition of the class (10.2),

$$\sum_{i,j,k=0}^{1} x_{ij}j_{jk}z_{ki} = \sum_{q=0}^{6} L_qL_q'L_q''.$$

Here

$$L_0L_0'L_0'' = (x_{00}+x_{11})(y_{00}+y_{11})(z_{00}+z_{11}),\ L_1L_1'L_1'' = (x_{10}+x_{11})y_{00}(z_{01}-z_{11}),$$

$$L_2L_2'L_2'' = x_{00}(y_{01}-y_{11})(z_{10}+z_{11}),\ L_3L_3'L_3'' = (-x_{00}+x_{10})(y_{00}+y_{01})z_{11},$$

$$L_4L_4'L_4'' = (x_{00}+x_{01})y_{11}(-z_{00}+z_{10}),\ L_5L_5'L_5'' = x_{11}(-y_{00}+y_{10})(z_{00}+z_{01}),$$

$$L_6L_6'L_6'' = (x_{01}-x_{11})(y_{10}+y_{11})z_{00}.$$

For the reverse transition from (10.2) to (2.3), consider the trilinear forms as linear forms in the z-variables and for all k,i equate the coefficients of z_{ki} on both sides of (10.2). If instead of that we equate the coefficients of x_{ij} or y_{jk} in (10.2), then we will obtain bilinear algorithms of the same rank M for the evaluation of YZ and ZX, respectively.

The latter observation has some interesting consequences in the case of rectangular MM. Whenever we have a bilinear algorithm for the problem (m,n,p), we may first represent it as a trilinear identity and then derive two <u>dual algorithms</u> for the problems (n,p,m) and (p,m,n) by equating the coefficients of two other groups of variables. All three dual algorithms have the same rank. Consequently all three dual problems (m,n,p),(n,p,m), (p,m,n) have the same rank, $\rho(m,n,p) = \rho(n,p,m) = \rho(p,m,n)$. The latter rank is also called the <u>rank of the trilinear form</u> Tr(XYZ) (it is also known as the <u>rank of the tensor of the coefficients of that trilinear form</u>) and is equal to the <u>minimum number of terms</u> in all possible decompositions (10.2) for the given trilinear form Tr(XYZ).

This implies Theorem 2.3 because we also can trivially reduce (m,n,p) to (n,m,p) just by interchanging the subscripts of the variables. Similarly Theorem 2.3 can be extended to <u>the case of λ-ranks of λ-algorithms for MM and for Disjoint MM</u>. ∎

Similarly to the representation (10.2) of bilinear algorithms for MM, we derive the trilinear versions of the general bilinear algorithms (2.4),(2.5) and of the bilinear λ-algorithms (6.7),

$$\sum_{\gamma=0}^{\Gamma-1} b_\gamma(X,Y)z_\gamma = \sum_{q=0}^{M-1} L_qL_q'L_q'',\ L_q'' = \sum_{\gamma=0}^{\Gamma-1} f''(\gamma,q)z_\gamma \text{ for all } q, \tag{10.3}$$

$$\lambda^d \sum_{\gamma=0}^{\Gamma-1} b_\gamma(X,Y)z_\gamma = \sum_{q=0}^{M-1} L_qL_q'L_q'' \text{ modulo } \lambda^{d+1},\ L_q'' = \sum_{\gamma=0}^{\Gamma-1} f''(\gamma,q,\lambda)z_\gamma \text{ for all } q. \tag{10.4}$$

L_q, L_q' for all q are defined by (2.5) in (10.3) and by (6.2) in (10.4); the coefficients f, f', f'' have the same meaning as in the bilinear algorithms (2.4),(2.5) and in the bilinear λ-algorithms (6.2),(6.7), respectively. Equating the coefficients of the z-variables on both sides of (10.3) or (10.4) we come back to (2.4),(2.5) or (6.7), respectively. The two dual algorithms associated with (2.4),(2.5) can be derived by equating the coefficients of the x-variables or of the y-variables on both sides of (10.3). Similarly two dual λ-algorithms associated with the λ-algorithm (6.2),(6.7) can be derived from (10.4). For instance, we have the trilinear identity (10.1) associated with the bilinear algorithm of Example 2.1 of Section 2. Equating the coefficients of x_0, x_1 on both sides of (10.1) we derive one of the two dual algorithms associated with the original algorithm of Example 2.1,

$$L_0L_0' = (y_0+y_1)z_0,\ L_1L_1' = y_1(-z_0+z_1),\ L_2L_2' = y_0(z_0+z_1),$$

$$y_0z_0+y_1z_1 = L_0L_0' + L_1L_1',\ y_0z_1-y_1z_0 = -L_0L_0' + L_2L_2'.$$

Again, all three dual algorithms have the same rank (3 in this example). The duality technique enables us to extend any successful bilinear algorithms to two new ones for two new problems, sometimes quite different from the original problem (see [K],[W80]). Historically the duality method was first applied to MM, and Theorem 2.3 was the first illustration of its power, see [P72],[HM].

We are going to work with algorithms and λ-algorithms for Disjoint MM. Thus here are the trilinear representation of a general bilinear algorithm of rank M for the evaluation over a ring F of S disjoint matrix products X(s)Y(s), s=0,1,...,S-1, and the trilinear representation of a general bilinear λ-algorithm of λ-rank M and degree d for the same problem.

$$\sum_{s=0}^{S-1} \mathrm{Tr}(X(s)Y(s)Z(s)) = \sum_{q=0}^{M-1} L_qL_q'L_q'', \tag{10.5}$$

$$\lambda^d \sum_{s=0}^{S-1} \mathrm{Tr}(X(s)Y(s)Z(s)) = \sum_{q=0}^{M-1} L_qL_q'L_q'' \text{ modulo } \lambda^{d+1}. \tag{10.6}$$

In (10.5),(10.6) L_q, L_q', and L_q'' are linear forms in the x-variables, in the y-variables, and in the z-variables, respectively, whose coefficients in (10.5) are from a ring F and in (10.6) are polynomials in λ over a ring F.

In particular, here are the trilinear representations of the bilinear algorithm (4.1),(4.2) and of the bilinear λ-algorithm (4.5),(4.6).

$$\mathrm{Tr}(XYZ) + \mathrm{Tr}(UVW) = \sum_{i,j,k} (x_{ij}y_{jk}z_{ki}+u_{jk}v_{ki}w_{ij}) =$$

$$\sum_{i,j,k} (x_{ij}+u_{jk})(y_{jk}+v_{ki})(z_{ki}+w_{ij}) - \tag{10.7}$$

$$\sum_{i,j} x_{ij} \sum_{k}(y_{jk}+v_{ki})w_{ij} - \sum_{j,k} u_{jk}y_{jk} \sum_{i}(z_{ki}+w_{ij}) -$$

$$\sum_{k,i} (\sum_{j}(x_{ij}+u_{jk}))v_{ki}z_{ki}.$$

$$\lambda(\mathrm{Tr}(XYZ) + \mathrm{Tr}(UVW)) = \lambda \sum_{i,j,k} (x_{ij}y_{jk}z_{ki}+u_{jk}v_{ki}w_{ij}) =$$

$$\sum_{i,j,k} (x_{ij}+u_{jk})(y_{jk}+\lambda v_{ki})(\lambda z_{ki}+w_{ij}) - \tag{10.8}$$

$$\sum_{i,j} x_{ij}\sum_{k}(y_{jk}+\lambda v_{ki})w_{ij} - \sum_{j,k} u_{jk}y_{jk}\sum_{i}(\lambda z_{ki}+w_{ij}) \text{ modulo } \lambda^2.$$

In the remainder of this section we will present a generalization of the equations (2.12) and Propositions 2.2,6.2, and 8.1. Such a generalization will enable us to represent the recursive algorithms for MM and Disjoint MM in the convenient form of trilinear identities. That representation will be used in the recursive constructions of Sections 14 and 16, which will define the λ-algorithms that will enable us to improve the exponent from 2.5161 to 2.496. Thus some readers might postpone reading the remainder of this section until they start reading Section 14. Those readers who feel happy already with the exponent 2.517 may omit the remainder of this section.

We will apply the concept of the Kronecker product of multidimensional tensors and of the associated polylinear forms. For the sake of completeness, we will first recall the general definition where we use the concept of tensor, see Definition 10.1 and the subsequent Remark 10.1. However, we will later give the independent and tensor-free definition of the Kronecker product in the special cases that we will actually need in our study of MM, that is, of the Kronecker product of two matrices and of the Kronecker products of two linear forms (see Definition 10.2).

<u>Definition 10.1</u>. The (d^*+d)-dimensional array (tensor) $\mathcal{T} = t^* \otimes t = [\mathcal{T}(i^*_0,\ldots,i^*_{d^*-1},i_0,\ldots,i_{d-1})] = [t^*(i^*_0,\ldots,i^*_{d^*-1}) \otimes t(i_0,\ldots,i_{d-1})]$ of the size $m^*_0 \times m^*_1 \times\ldots\times m^*_{d^*-1} \times m_0 \times m_1 \times\ldots\times m_{d-1}$ is called <u>the Kronecker product or the direct product of the two tensors (arrays)</u> $t^* = [t^*(i^*_0,\ldots,i^*_{d^*-1})]$ and $t = [t(i_0,\ldots,i_{d-1})]$ of dimensions d^* and d, respectively, and of sizes $m^*_0 \times m^*_1 \times\ldots\times m^*_{d^*-1}$ and $m_0 \times m_1 \times\ldots\times m_{d-1}$, respectively. Here the natural parameters

$i_0^*,\ldots,i_{d^*-1}^*, i_0,\ldots,i_{d-1}$ range from 0 to $m_0^*-1,\ldots,m_{d^*-1}^*-1, m_0-1,\ldots m_{d-1}$, respectively. The (d^*+d)-linear form $D^*(t^*) \otimes D(t)$ associated with the tensor $t^* \otimes t$ is called the *Kronecker product of the* d^*-*linear form* $D^*(t^*)$ *associated with the tensor* t^* *and of the* d-*linear form* D(t) *associated with the tensor* t.

It can be immediately verified that

$$D^*(t^*) \otimes D(t) = \Sigma^* \ \Sigma\, \tau(i_0^*,\ldots,i_{d^*-1}^*, i_0,\ldots,i_{d-1}) u_0 u_1 \ldots u_{d^*-1} v_0 v_1 \ldots v_{d-1} =$$

$$D^*(t^*) D(t) = \Sigma^* \ t^*(i_0^*,\ldots,i_{d^*-1}^*) u_0 u_1 \ldots u_{d^*-1} \ \Sigma\, t(i_0,\ldots,i_{d-1}) v_0 v_1 \ldots v_{d-1}$$

(where Σ^* designates the sum in $i_0^*, i_1^*,\ldots,i_{d^*-1}^*$ and Σ designates the sum in $i_0, i_1,\ldots,i_{d-1}$) so that the Kronecker product of two polylinear forms coincide with their conventional product (and *actually this fact can be used as the definition of the Kronecker product of the tensors of the coefficients of those polylinear forms*).

Remark 10.1. Hereafter we will slightly modify Definition 10.1 by assuming that *the resulting* (d^*+d)-*dimensional tensor is defined up to within its equivalent transformation into tensors of smaller dimension*. (It is assumed that *similarly the associated polylinear forms are considered equivalent*.) For simplicity, we will explicitly show the equivalent transformation only in the case where $d^*+d = 2$. Specifically, any $m \times n$ matrix $\tau = [\tau(i,j), i=0,1,\ldots,m-1; j=0,1,\ldots,n-1]$, which is a 2-dimensional tensor, can be represented also as the vector $\underline{t} = [t(s), s=0,1,\ldots,mn-1]$ where $t(i+mj) = \tau(i,j)$ for $i=0,1,\ldots,m-1$; $j=0,1,\ldots,n-1$; $\underline{t}$ is a 1-dimensional tensor. (Note the possibility of the converse transformation of the vector $\underline{t}$ into the matrix τ. Similarly for the tensors of higher dimensions. Such an equivalent transformation of the (d^*+d)-dimensional tensors into the tensors of smaller dimensions D induces the equivalent transformation of the associated (d^*+d)-linear forms into D-linear forms. In particular, the above reduction of the matrix τ to the vector $\underline{t}$ induces the transformation of the bilinear form $b(U,V) = \sum_{i,j} \tau(i,j) u_i v_j$ associated with the matrix τ into the linear form

$l(W) = \sum_{s=0}^{mn-1} t(s) w_s = \sum_{i,j} \tau(i,j) w_{i+jm}$ associated with the vector $\underline{t}$. Note that the same transformation of $b(U,V)$ into $l(W)$ can be defined by the substitution of variables

$$w_{i+jm} = u_i v_j \text{ for all } i,j. \tag{10.9}$$

(Similar substitutions define the transformation of polylinear forms induced by the reduction of the associated tensors in the general case.)

The recursive bilinear algorithms can be conveniently defined via the Kronecker product of the associated trilinear forms with the subsequent reduction of the dimension 6 of the result back to 3 by three substitutions of variables of the type (10.9). To be formally independent of the concept of tensors, even though that concept is not hard to use, as we could have just seen, we will now give an independent and tensor-free definition of the Kronecker products of pairs of matrices, as well as of pairs of linear forms, with the subsequent reduction of the dimension from 2 to 1. These will be the only two cases that we will need in our study of MM. Some readers may prefer to use Definition 10.1 and Remark 10.1 where the concept of the Kronecker product appears more naturally.

Definition 10.2. The $(m^*m) \times (n^*n)$ matrix $W = U \otimes V = [w(i^*+im^*, j^*+jn^*)$, $i^*=0,1,\ldots,m^*-1$; $i=0,1,\ldots,m-1$; $j^*=0,1,\ldots,n^*-1$; $j=0,1,\ldots,n-1]$ where $w(i^*+im^*, j^*+jn^*) = w(i^*,i,j^*,j) = u(i^*,j^*)v(i,j)$ for all i^*,i,j^*,j is called the Kronecker product or the direct product of the two matrices $U = [u(i^*,j^*)$, $i^*=0,1,\ldots,m^*-1$; $j^*=0,1,\ldots,n^*-1]$ of the size $m^* \times n^*$ and $V=[v(i,j)$, $i=0,1,\ldots,m-1$; $j=0,1,\ldots,n-1]$ of the size $m \times n$. (To avoid using 4-tuple subscripts, we have raised all subscripts i^*,i,j^*,j and will handle such a problem similarly in the sequel.) If $n^*=n=1$ this defines an (m^*m)-vector as the Kronecker product of an m^*-vector and of an m-vector. The Kronecker product of two linear forms, $L^* = \sum_\mu f^*(\mu)u_\mu$ (in the u-variables) and $L = \sum_\eta f(\eta)v_\eta$ (in the v-variables) is defined as the linear form $L^* \otimes L = \sum_{\mu,\eta} f^*(\mu)f(\eta)w_{\mu\eta}$ (in the w-variables) whose vector of coefficients equals $f^* \otimes f$. Here f^* and f are the two vectors of the coefficients of L^* and L, respectively.

Obviously, if we substitute the products $u_\mu v_\eta$ for $w_{\mu\eta}$ for all μ,η, then $L^* \otimes L$ becomes identically equal to L^*L (compare (10.9)). The latter observation immediately leads to the following result.

Proposition 10.1. Let d^*,d be arbitrary natural numbers. Let two bilinear λ-algorithms B^* and B for MM or Disjoint MM be defined by the two following trilinear identities,

$$\left.\begin{aligned} \lambda^{d^*}\sum_{s^*=0}^{S^*-1} \mathrm{Tr}(X^*(s^*)Y^*(s^*)Z^*(s^*)) &= \sum_{q^*=0}^{M^*-1} L^*_{q^*}L^{*\prime}_{q^*}L^{*\prime\prime}_{q^*} \text{ modulo } \lambda^{d^*+1}, \\ \lambda^{d}\sum_{s=0}^{S-1} \mathrm{Tr}(X(s)Y(s)Z(s)) &= \sum_{q=0}^{M-1} L_qL'_qL''_q \text{ modulo } \lambda^{d+1}. \end{aligned}\right\} \quad (10.10)$$

(In particular one or both of these λ-algorithms can be bilinear algorithms if d^* and/or d equal zero.) Then the following trilinear identity defines the bilinear λ-algorithm $B^* \otimes B$ (which is called the product of the two given λ-algorithms, compare Propositions 2.2,6.2,8.1).

$$\lambda^{d^*+d} \sum_{s^*=0}^{S^*-1} \sum_{s=0}^{S-1} \mathrm{Tr}((X^*(s^*) \otimes X(s))(Y^*(s^*) \otimes Y(s))(Z^*(s) \otimes Z(s))) =$$

$$\sum_{q^*=0}^{M^*-1} \sum_{q=0}^{M-1} (L_q^* \otimes L_q)(L_q^{*'} \otimes L_q')(L_q^{*''} \otimes L_q'') \text{ modulo } \lambda^{d^*+d+1}. \quad (10.11)$$

Proof. Let at first $d^*=d=0$. Then the identity (10.11) is just the conventional product of the two identities (10.10) where the new variables u,v,w substitute for the products of the pairs of the variables x^*x, y^*y, z^*z, respectively; compare (10.9),

$$\left.\begin{array}{c} u(i^*,i,j^*,j) = x^*(i^*,j^*)x(i,j), v(j^*,j,k^*,k) = y(j^*,k^*)y(j,k), \\ w(k^*,k,i^*,i) = z^*(k^*,i^*)z(k,i) \text{ for all } i^*,j^*,k^*,i,j,k. \end{array}\right\} \quad (10.12)$$

In the general case where d^*,d are arbitrary natural numbers, the same proof is applied, with the additional observation that, as follows from (10.10), $\lambda^{d^*+1}P^* \sum_q L_q L_q' L_q'' = 0$ modulo λ^{d^*+d+1}, $\lambda^{d+1}P \sum_{q^*} L_{q^*}^* L_{q^*}^{*'} L_{q^*}^{*''} = 0$ modulo λ^{d^*+d+1} for all polynomials P^*,P in λ. This means that the terms, which are zeroes modulo λ^{d^*+1} and λ^{d+1} in the first and second identities of (10.10), respectively, will not influence the product of those two identities modulo λ^{d^*+d+1}. ∎

Remark 10.2. Proposition 10.1 implies the equations (2.12) and the three Propositions 2.2, 6.2, and 8.1, which it generalizes. Therefore our proof of Proposition 10.1 is a unified proof of all of those results. On the other hand, using the straightforward extension of Definition 10.2 to the case of the Kronecker products of trilinear forms with the subsequent reduction of the 6-linear forms back to trilinear ones (compare Definition 10.1 and Remark 10.1) we can define the general recursive construction that starts with the trilinear representation of a bilinear algorithm or of a bilinear λ-algorithm for an arbitrary bilinear problem.

11. Trilinear Aggregating and Some Efficient Basis Designs

If we rewrite bilinear λ-algorithms as trilinear decompositions, then the numbers of terms $L_q L_q' L_q''$ modulo λ^{d+1} on the right sides will equal the λ-ranks of those algorithms (similarly for the ranks of conventional bilinear algorithms).

Therefore our objective can be restated as the search for the trilinear decompositions with fewer terms. Let us reexamine decompositions (10.7),(10.8) from that point of view.

At first represent the principal terms $x_{ij}y_{jk}z_{ki}$ and $u_{jk}v_{ki}w_{ij}$ in the two following tables, each consisting in one line.

Table 11.1

x_{ij}	y_{jk}	z_{ki}

Table 11.2

u_{jk}	v_{ki}	w_{ij}

Next aggregate the two tables into the following one.

Table 11.3

x_{ij}	y_{jk}	z_{ki}
u_{jk}	v_{ki}	w_{ij}

Sum the pairs of entries in each column of Table 11.3 and multiply the three sums together. The resulting aggregate $(x_{ij}+u_{jk})(y_{jk}+v_{ki})(z_{ki}+w_{ij})$ is a substitution for the sum of the two principal terms $x_{ij}y_{jk}z_{ki}$ and $u_{jk}v_{ki}w_{ij}$. The latter sum differs from the aggregate by the sum of six correction terms, that is, of the six cross-products of Table 11.3, $x_{ij}y_{jk}w_{ij}$, $x_{ij}v_{ki}z_{ki}$, $x_{ij}v_{ki}w_{ij}$, $u_{jk}y_{jk}z_{ki}$, $u_{jk}y_{jk}w_{ij}$, $u_{jk}v_{ki}z_{ki}$. (Our aggregating of two tables into one may remind of the well-known method of linear aggregating, compare [MP].) The main point now is that we deal with mnp pairs of principal terms associated with all mnp existing triplets i,j,k. Therefore Table 11.3 for all i,j,k defines mnp aggregates. The key idea of our trilinear design is to correct the discrepancy between the sums of the mnp aggregates and of the 2mnp principal terms by using just a few (only mn+np+pm) correction terms. Then the total number of terms will be reduced from the 2mnp principal terms to the mnp+mn+np+pm terms on the right side of (10.7). Table 11.3 defines the algorithm (10.7) up to regrouping the correction terms. (For this particular table, the optimum regrouping is straightforward.)

A similar table is associated with the algorithm (10.8) and also defines that algorithm up to the straightforward regrouping of correction terms.

Aggregating Table 11.4

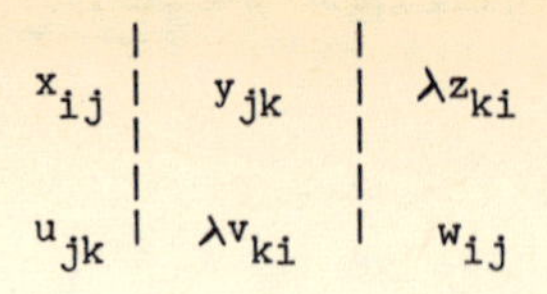

Again, the two products of the three entries of each row define two principal terms. Their aggregate is the product of the three sums, $x_{ij}+u_{jk}$, $y_{jk}+\lambda v_{ki}$, and $\lambda z_{ki}+w_{ij}$. For each triplet i,j,k the correction term $\lambda^2(x_{ij}+u_{jk})v_{ki}z_{ki}$ vanishes modulo λ^2. Thus comparing with Table 11.3 the total number of terms has been reduced.

In the sequel we will use S X 3 aggregating tables of this kind for different values S as a means of compact but equivalent (up to the regrouping of the correction terms) representation of efficient bilinear algorithms and λ-algorithms for MM and Disjoint MM. In particular, we will see in Sections 14 and 16 how the recursive algorithms and λ-algorithms can be defined using the Kronecker products of the aggregating tables associated with the basis algorithms.

12. A Further Example of Trilinear Aggregating and Its Refinement via a Linear Transformation of Variables

In this section we will demonstrate how a rather simple modification of the design (4.5),(4.6) leads to a substantial decrease of the exponent.

At first recall that the bound 2.67 was obtained from the design (4.5),(4.6) in the case where n=1. In that case j is fixed, j=0, and Table 11.4 can be rewritten and then modified as follows.

Aggregating Table 12.1

x_{i0}	y_{0k}	λz_{ki}
u_{0k}	λv_{ki}	w_{i0}

Aggregating Table 12.2

x_{i0}	y_{0k}	$\lambda^2 z_{ki}$
λu_{0k}	λv_{ki}	w_{i0}

Table 12.2 has been obtained from Table 12.1 (by multiplying u_{0k} and λz_{ki} by λ). The grouping of the correction terms remains the same as in the case of Tables 11.3 and 11.4.

The second line of Table 12.2 defines the principal terms $\lambda^2 u_{0k} v_{ki} w_{i0}$ for all k,i. Their sum represents the problem $\lambda^2 \circledcirc (1,p,m)$. Consider also the problem $\lambda^2 \circledcirc (1,mp,1)$, which can be represented by the terms $\lambda^2 u_{0,i+mk} v_{i+mk,0} w_{00}$. Simplify the notation by rewriting them as $\lambda^2 u_{0,k,i} v_{k,i,0} w_{00}$ and represent them as the principal terms of the following one-line table.

Table 12.3

$\lambda u_{0,k,i}$	$\lambda v_{k,i,0}$	w_{00}

Aggregate Table 12.3 and the following one-line table.

Table 12.4

x_{i0}	y_{0k}	$\lambda^2 z_{ki}$

Obtain the following table as the result.

Table 12.5

x_{i0}	y_{0k}	$\lambda^2 z_{ki}$
$\lambda u_{0,k,i}$	$\lambda v_{k,i,0}$	w_{00}

Table 12.5 defines the following λ-algorithm, which served as <u>the main design of</u> [S81] (our tables may help to visualize the origin of that design, which appears ad hoc in [S81]).

$$\lambda^2 \sum_{i,k} (x_{i0} y_{0k} z_{ki} + u_{0,k,i} v_{k,i,0} w_{00}) =$$

$$\sum_{i,k} (x_{i0} + \lambda u_{0,k,i})(y_{0k} + \lambda v_{k,i,0})(\lambda^2 z_{ki} + w_{00}) - \tag{12.1}$$

$$\sum_i x_{i0} \sum_k y_{0k} w_{00} - \lambda \sum_i x_{i0} (\sum_k v_{k,i,0}) w_{00} -$$

$$\lambda \sum_k (\sum_i u_{0,k,i}) y_{0k} w_{00} \text{ modulo } \lambda^3.$$

Here i and k range from 0 to $m-1$ and $p-1$, respectively. $\sum_{i,k} u_{0,k,i}v_{k,i,0}w_{00}$ represents $1 \times p \times 1$ MM. (To see this, designate that $u_{0,i+mk} = u_{0,k,i}, v_{i+mk,0} = v_{k,i,0}$.)

Next assume that not all $u_{0,k,i}, v_{k,i,0}$ are indeterminates but that those variables satisfy the following linear equations,

$$\sum_{i=0}^{m-1} u_{0,k,i} = 0, \; u_{0,k,0} = -\sum_{i=1}^{m-1} u_{0,k,i} \text{ for all } k. \tag{12.2}$$

$$\sum_{k=0}^{p-1} v_{k,i,0} = 0, \; v_{0,i,0} = -\sum_{k=1}^{p-1} v_{k,i,0} \text{ for all } i. \tag{12.3}$$

Then the number of correction terms in (12.1) decreases to 1. We also introduce the following complementary equations (which complement (12.2) and (12.3)).

$$u_{0,0,i} = 0 \text{ for all } i. \tag{12.4}$$

$$v_{k,0,0} = 0 \text{ for all } k. \tag{12.5}$$

Substitute the expressions for $u_{0,k,0}, v_{0,i,0}, u_{0,0,i}, v_{k,0,0}$ from (12.2)-(12.5) into (12.1) and assume that all variables u,v,w,x,y,z are indeterminates except for those variables $u_{0,k,i}$ and $v_{k,i,0}$ where either $i=0$, or $k=0$, or $i=k=0$. (The latter variables are defined by the equations (12.2)-(12.5). They either equal 0 or are expressed as some linear combinations of $v_{k,i,0}$ and $u_{0,k,i}$ where $i=1,\dots,m-1$ and $k=1,\dots,p-1$.) Then it is immediately verified that the left-hand side of (12.1) turns into the trilinear form associated with the problem $\lambda^2 \odot ((m,1,p) \oplus (1,(m-1)(n-1),1))$. Therefore, the trilinear identity is represented by the following mapping (λ-algorithm),

$$\lambda^2 \odot ((m,1,p) \oplus (1,(m-1)(p-1),1)) \leftarrow (mp+1) \odot (1,1,1). \tag{12.6}$$

Application of Theorem 7.2 to the mapping (12.6) for $m=p=4$ yields the bound

$$\omega < 2.548 \tag{12.7}$$

(over arbitrary ring of constants F). Note that the cancelling of m+n-1 (out of m+n) correction terms has been achieved by imposing some linear equations on the variables, which reduced the dimension of the problem also by m+n-1. It turns out that such a trade-off (the reduction of both values, that is, of the rank and of the dimension, by the same quantity m+n-1) always implies the reduction of the real root τ of the associated equation (7.2) and consequently of the exponent ω. Of course, we could also cancel the remaining correction term $\sum_i x_{i0} \sum_k y_{0k} w_{00}$ by imposing an additional equation, say, $\sum_i x_{i0} = 0$, but this would not be efficient due to the resulting decrease of the dimension of the problem by p-1. In general, we always should compare the influence of the imposed equations onto both values (the dimension and the number of correction terms). Actually in the better designs some equations are imposed and most of the correction terms (but not all of them) are canceled.

The comparative examination of different designs convinces that the bound (12.7) is the best one that can be obtained via aggregating the pairs of principal terms. In the search for the asymptotically faster methods for MM we may and will try to aggregate more principal terms. This will require to use relatively more correction terms or to impose more equations (which will cancel most of or some of those terms but simultaneously will require to reduce the dimension of the problem). Due to the contradictory influence of the latter two factors on the associated exponent, it is not clear a priori whether we may finally reduce the total number of terms and the exponent by aggregating more principal terms. Actually, we will be able to push the reduction a little further by aggregating more terms in a special way and by using some special canceling procedures, see Remark 14.2 in Section 14.

13. Aggregating the Triplets of Principal Terms

In this section we will present an approach that leads to the exponent 2.5161. The techniques used will also help us later in designing fast algorithms for N X N MM for moderate N (see Section 31). In Sections 14, 15, and 16, which can be read independently of this section, we will present another approach that leads to the exponent 2.496 (but does not help for N X N MM unless N is enormously large).

Next we will show how to improve the estimate (12.7) and to derive the bound $\omega < 2.522$ by aggregating some triplets of the principal terms.

Our design will rely on the two following tables.

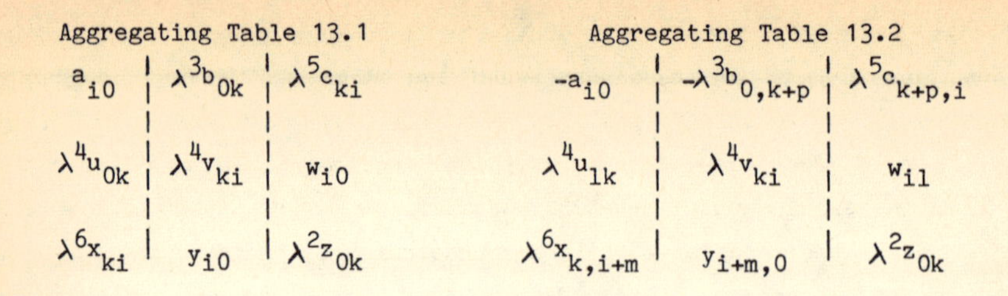

Aggregating Table 13.1

a_{i0}	$\lambda^3 b_{0k}$	$\lambda^5 c_{ki}$
$\lambda^4 u_{0k}$	$\lambda^4 v_{ki}$	w_{i0}
$\lambda^6 x_{ki}$	y_{i0}	$\lambda^2 z_{0k}$

Aggregating Table 13.2

$-a_{i0}$	$-\lambda^3 b_{0,k+p}$	$\lambda^5 c_{k+p,i}$
$\lambda^4 u_{1k}$	$\lambda^4 v_{ki}$	w_{i1}
$\lambda^6 x_{k,i+m}$	$y_{i+m,0}$	$\lambda^2 z_{0k}$

Starting with Tables 13.1,13.2, we will come to the following result.

Theorem 13.1. Let $t = (m-1,1,2p-2) \oplus (2,p-1,m-1) \oplus (p-1,2m-2,1)$, $M = 2m(p+1)$, $m>1, p>1$. Let F be an arbitrary ring of constants. Then there exists a bilinear λ-algorithm over F represented by the following mapping, $\lambda^8 \otimes t \leftarrow M \otimes (1,1,1)$.

The application of Theorem 7.2 to the above λ-algorithm for $m=6$, $p=12$ yields the bound $\omega_F \leq 3 \log 52/\log 110 < 2.522$ over all rings F.

Proof. Aggregate the triplets of principal terms using Tables 13.1,13.2 for $i=0,1,\dots,m-1$; $k=0,1,\dots,p-1$. Then sum the aggregates of both tables together and sum the resulting terms in all i,k. Note that the correction terms $\lambda^6 a_{i0} v_{ki} z_{0k}$ and $-\lambda^6 a_{i0} v_{ki} z_{0k}$ cancel themselves out for all i,k and that the reduction modulo λ^9 cancels several other correction terms. We would like to cancel all remaining correction terms or at least most of them by imposing certain linear equations onto the variables a,b,c,u,v,w,x,y,z. This leads us to the following equations.

$$\sum_i a_{i0} = \sum_i w_{i0} = \sum_i w_{i1} = \sum_i y_{i0} = \sum_i y_{i+m,0} = 0, \tag{13.1}$$

$$\left.\begin{array}{l} \sum_k b_{0k} = \sum_k b_{0,k+p} = \sum_k u_{0k} = \sum_k u_{1k} = \sum_k z_{0k} = 0, \\ \sum_k c_{ki} = \sum_k c_{k+p,i} = \sum_k v_{ki} = \sum_k x_{ki} = \sum_k x_{k,i+m} = 0 \text{ for all } i. \end{array}\right\} \tag{13.2}$$

Here and hereafter $\sum_i$ and $\sum_k$ designate the sums in i and k ranging from 0 to m-1 and to p-1, respectively. Under the equations (13.1),(13.2), we are left with the sum of only 2m correction terms,

$$\sigma = p \sum_i a_{i0}(-y_{i0} w_{i0} + y_{i+m,0} w_{i1}) \text{ modulo } \lambda^9. \tag{13.3}$$

This design is already not bad but we will certainly improve it if we cancel as many correction terms by imposing fewer equations on the variables. To achieve that goal, we will relax some of the equations (13.2) and will add some correction terms to σ so that, on the one hand, the resulting sum σ^* plus the sum of all aggregates will be equal to the sum of all principal terms under the remaining equations of (13.1),(13.2) and, on the other hand, the terms added to σ will not change the rank of the trilinear form σ of (13.3), that is, we will still be able to rewrite the resulting trilinear form σ^* as a sum of 2m terms.

Specifically let us proceed as follows. Impose all equations (13.1) on the variables a,y,w. Impose only the equations of the following subsystem of (13.2) on all other variables.

$$\sum_k u_{0k} = \sum_k u_{1k} = 0,\ \sum_k c_{ki} = \sum_k c_{k+p,i} = \sum_k x_{ki} = \sum_k x_{k,i+m} = 0 \text{ for all } i. \quad (13.4)$$

Note that several equations of the system (13.2) have been omitted. All of the 2m terms of σ, as well as some other correction terms, remain uncanceled in spite of imposing the equations (13.1),(13.4). Next we will show that all of those uncanceled terms can be aggregated into the two following aggregating tables for $i=0,1,\ldots,m-1$.

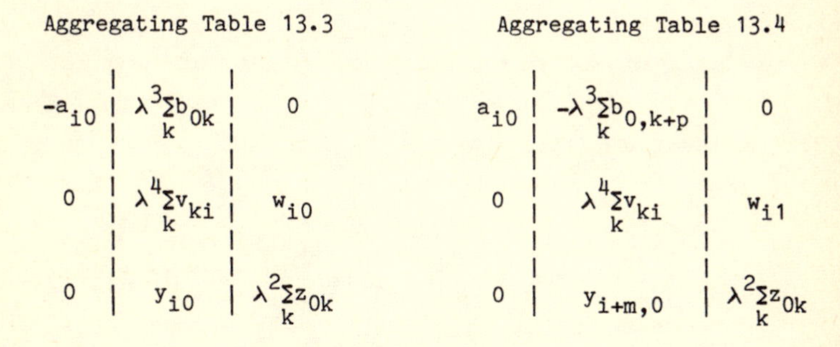

Aggregating Table 13.3

$-a_{i0}$	$\lambda^3\sum_k b_{0k}$	0
0	$\lambda^4\sum_k v_{ki}$	w_{i0}
0	y_{i0}	$\lambda^2\sum_k z_{0k}$

Aggregating Table 13.4

a_{i0}	$-\lambda^3\sum_k b_{0,k+p}$	0
0	$\lambda^4\sum_k v_{ki}$	w_{i1}
0	$y_{i+m,0}$	$\lambda^2\sum_k z_{0k}$

Indeed, it is easy to check that all correction terms of Tables 13.1 and 13.2 that contain $c_{ki}, c_{k+p,i}, u_{0k}, u_{1k}, x_{ki}, x_{k,i+m}$ disappear in the result of their summation in i and k and the reduction modulo λ^9 provided that (13.1) and (13.4) hold. Therefore it suffices to add the sum in all i and k of the aggregates of the following two tables in order to cancel the remaining correction terms.

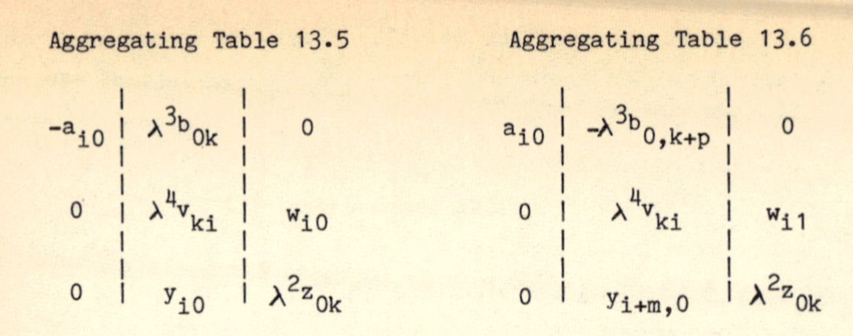

Aggregating Table 13.5		
$-a_{i0}$	$\lambda^3 b_{0k}$	0
0	$\lambda^4 v_{ki}$	w_{i0}
0	y_{i0}	$\lambda^2 z_{0k}$

Aggregating Table 13.6		
a_{i0}	$-\lambda^3 b_{0,k+p}$	0
0	$\lambda^4 v_{ki}$	w_{i1}
0	$y_{i+m,0}$	$\lambda^2 z_{0k}$

It remains to note that the sum in i and k of all aggregates of Tables 13.5 and 13.6 coincide with the sum in i of all aggregates of Tables 13.3 and 13.4. The sums of the terms $-\lambda^5 \sum_i a_{i0} \sum_k b_{0k} z_{0k}$, $-\lambda^5 \sum_i a_{i0} \sum_k b_{0,k+p} z_{0k}$, $-\lambda^5 \sum_i a_{i0} \sum_k b_{0k} \sum_k z_{0k}$, and $-\lambda^5 \sum_i a_{i0} \sum_k b_{0,k+p} \sum_k z_{0k}$ generated by Tables 13.3, 13.4, 13.5, and 13.6, respectively, are canceled since $\sum_i a_{i0} = 0$, see (13.1), and the pairs of the undesirable terms $-\lambda^6 a_{i0} v_{ki} z_{0k}$ and $\lambda^6 a_{i0} v_{ki} z_{0k}$, $-\lambda^6 a_{i0} \sum_k v_{ki} \sum_k z_{0k}$ and $\lambda^6 a_{i0} \sum_k v_{ki} \sum_k z_{0k}$ generated by the pairs of Tables 13.5 and 13.6, 13.3 and 13.4, respectively, cancel themselves out for all i when the two aggregates of each pair of tables are summed together.

Therefore we are left with only 2mp aggregates of Tables 13.1, 13.2 for all i,k and with 2m aggregates of Tables 13.3, 13.4 for all i, that is, with a total of 2mp+2m terms. That means that we have devised a bilinear λ-algorithm of λ-rank M=2mp+2m for the bilinear problem represented by the sum of all principal terms of Tables 13.1 and 13.2 over all i,k under the linear equations (13.1),(13.4). The sum of the principal terms of those tables in all i,k is equal to the following trilinear form,

$$T = \lambda^8 \sum_{i,k} (a_{i0}(b_{0k} c_{ki} + b_{0,k+p} c_{k+p,i}) + u_{0k} v_{ki} w_{i0} + \tag{13.5}$$

$$u_{1k} v_{ki} w_{i1} + (x_{ki} y_{i0} + x_{k,i+m} y_{i+m,0}) z_{0k}).$$

This would have defined the problem $\lambda^8 \odot ((m,1,2p) \oplus (2,p,m) \oplus (p,2m,1))$ if all variables a,b,c,u,v,w,x,y,z had been indeterminates. Actually they are not indeterminates, due to the equations (13.1),(13.4), and the trilinear form T is associated with another problem of Disjoint MM, of a smaller dimension. To determine that problem, we will follow the line of the previous section. Rewrite the equations (13.1),(13.4) as follows,

$$a_{00} = -\sum_i^* a_{i0},\ w_{00} = -\sum_i^* w_{i0},\ w_{01} = -\sum_i^* w_{i1},\ y_{00} = -\sum_i^* y_{i0},\ y_{m0} = -\sum_i^* y_{i+m,0},$$

$$u_{00} = -\sum_k^* u_{0k},\ u_{10} = -\sum_k^* u_{1k},$$

$$c_{0i} = -\sum_k^* c_{ki}, c_{pi} = -\sum_k^* c_{k+p,i}, x_{0i} = -\sum_k^* x_{ki}, x_{0,i+m} = -\sum_k^* x_{k,i+m} \text{ for all } i.$$

Here the symbols $\sum_i^*$ and $\sum_k^*$ designate the sums in i and k ranging from 1 to m-1 and to p-1, respectively. Assume that only the variables on the left sides, that is, $c_{0i}, c_{pi}, x_{0i}, x_{0,i+m}$ for all i and $a_{00}, w_{00}, w_{01}, y_{00}, y_{m0}, u_{00}, u_{10}$, are not indeterminates. Eliminate all of those variables from the trilinear form T of (13.5) by substituting the variables expressed by the following complementary equations into the trilinear form T (compare (12.2)-(12.5)).

$$c_{k0} = c_{k+p,0} = v_{k0} = x_{k0} = x_{k,m} = 0 \text{ for all } k,\ v_{0i} = 0 \text{ for all } i,\ b_{00} = b_{0p} = z_{00} = 0.$$

Then the trilinear form T of (13.5) turns into the following trilinear form in the unbound variables, (indeterminates),

$$T = \lambda^8 \sum_i^* \sum_k^* (a_{i0}(b_{0k}c_{ki} + b_{0,k+p}c_{k+p,i}) + u_{0k}v_{ki}w_{i0} + u_{1k}v_{ki}w_{i1} +$$

$$(x_{ki}y_{i0} + x_{k,i+m}y_{i+m,0})z_{0k}).$$

The expression under the summation symbol is the same as in (13.5). However, i and k range in a smaller interval, that is, from 1 to m-1 and p-1, respectively. Since all remaining variables are indeterminates, the trilinear form T is associated with the problem of the desired size,

$$\lambda^8 \otimes ((m-1,1,2p-2) \oplus (2,p-1,m-1) \oplus (p-1,2m-2,1)).$$

We have already proven that the λ-rank of that trilinear form T is equal to at most M=2m(p+1). ■

<u>Remark 13.1</u>. The bilinear λ-algorithm of Theorem 13.1 first appeared in the Remark on p. 37 of [P79]. The presented above derivation of that λ-algorithm is substantially simpler than the original way due to the linear transformation of variables (13.1),(13.4), which was not used in [P79]. Still the design of [P79] is a great simplification of similar constructions of [P78] and [P80] (called 3-Procedure) due to the benefits of the reduction modulo λ^9. (In [P78],[P80] similar methods of aggregating were applied for designing conventional algorithms.) Also the cited construction from [P79] was historically the first example that disproved the

extension of Strassen's conjecture to the case of λ-algorithms or equivalently to the case where the constants are taken from the ring of polynomials in λ modulo λ^{d+1}, see Theorem 7.1 in Section 7. (Indeed, the well known basic active substitution techniques, see Theorem 36.1 in Section 36, immediately lead to the lower bound maximum {mn,np,pm} on the λ-rank of the problem (m,n,p). This would imply the lower bound 5(m-1)(p-1) on the λ-rank of the problem $\lambda^8 \odot ((m-1,1,2p-2) \oplus (2,p-1,m-1) \oplus (p-1,2m-2,1))$ if Strassen's conjecture had held true over λ-algorithms. Since the λ-rank of that problem has been shown to be not greater than M=2m(p+1), which is less than 5(m-1)(p-1), say, for m=p=5, we arrive at a formal disproval of the extended Strassen's conjecture.) The design of this counter-example, as well as of all other known ones, relies on the techniques of trilinear aggregating. A. Schönhage was the first who fully appreciated and thoroughly investigated and exploited the deep relationship between the trilinear aggregating techniques and Disjoint MM, see [S81]. He certainly deserves full credit for that. However, for the sake of historical accuracy, we would like to correct his misleading statement that our λ-algorithm of Theorem 13.1 has been designed "soon after" (we cite [S81], p. 435) he sketched the proof of Theorem 7.2 at the Oberwolfach Conference on October 24, 1979. Actually the algorithm was published in October 1979 (see Remark on p. 37 of [P79]), was presented at the same conference on October 23, 1979, and was devised well before, rather than "soon after" that conference, as well as, more generally, several fast MM algorithms and λ-algorithms have been designed using trilinear aggregating well before (and some of them have been even published well before) Schönhage's sketchy presentation of Theorem 7.2 in October 1979, see [P72],[P78],[P79]. Surely the misleading statement by A. Schönhage was not his accidental error (he was well informed about the subject) but rather a means in his competition for the leadership in the acceleration of MM in 1979. Being an organizer of the Oberwolfach conference in 1979 he managed to exclude any remark about the exponent 2.522 from the official report on the conference leaving himself with "the world record 2.548" (we cite that official report, compare the appendix to this book). A. Schönhage continued the competition at the Oberwolfach conferences in 1981 and 1983 in a similar way. Being responsible for the program of these two conferences, he succeeded in cutting the time for the lecture of the present author by 25% at the last moment before that lecture in 1981 and in excluding the lecture of that author from the official program of the conference in 1983. (He could not, however, stop that lecture, which was given unofficially upon the request from several participants of the conference.)

It is possible to improve the exponent 2.522 to 2.5161 by applying a modification of the recursive construction of Theorem 7.2 to the presented above design. In the remainder of this section we will indicate the direction for such a modification. At first note that for each i the two aggregates of Tables 13.3 and 13.4 have a common factor a_{i0}, so that we may represent those two aggregates as the products $A_{00}B_{00}C_{00}$ and $A_{00}B_{01}C_{10}$ where

$$A_{00} = -a_{i0},\ B_{00} = \lambda^3 \sum_k b_{0k} + \lambda^4 \sum_k v_{ki} + y_{i0},\ C_{00} = w_{i0} + \lambda^2 \sum_k z_{0k},$$

$$B_{01} = \lambda^3 \sum_k b_{0,k+p} - \lambda^4 \sum_k v_{ki} - y_{i+m,0},\ C_{10} = w_{i1} + \lambda^2 \sum_k z_{0k}.$$

Then for each i the sum of those two aggregates $A_{00}(B_{00}C_{00} + B_{01}C_{10})$ represents the problem (1,1,2), that is, 1 X 1 X 2 MM. Therefore the sum of m disjoint pairs of such aggregates represents the problem $m \odot (1,1,2)$ rather than $2m \odot (1,1,1)$. That means that we have actually constructed the bilinear λ-algorithm

$$\left.\begin{aligned} \lambda^8 \odot t &\leftarrow t_0, \\ t &= (m-1,1,2p-2) \oplus (2,p-1,m-1) \oplus (p-1,2m-2,1), \\ t_0 &= 2mp \odot (1,1,1) \oplus m \odot (1,1,2). \end{aligned}\right\} \quad (13.6)$$

The above algorithm may seem to give no improvement over Theorem 13.1 because the problem t_0 on the righthand side has rank $M=2m(p+1)$. Therefore the best we can do for the solution of the problem t_0 is just to reduce it back to $M \odot (1,1,1)$. However, recall that already for the problem (2,2,2) we have an algorithm of rank 7 (see Section 1). The terms (2,2,2) will appear on the right side if we multiply together the λ-algorithm (13.6) and the two dual λ-algorithms,

$$\lambda^8 \odot t' \leftarrow t_0',\ \lambda^8 \odot t'' \leftarrow t_0''. \quad (13.7)$$

Here

$$\left.\begin{aligned} t_0' &= 2mp \odot (1,1,1) \oplus m \odot (1,2,1),\ t_0'' = 2mp \odot (1,1,1) \oplus m \odot (2,1,1), \\ t' &= (1,2p-2,m-1) \oplus (p-1,m-1,2) \oplus (2m-2,1,p-1), \\ t'' &= (2p-2,m-1,1) \oplus (m-1,2,p-1) \oplus (1,p-1,2m-2). \end{aligned}\right\} \quad (13.8)$$

The resulting product is the (bilinear) λ-algorithm

$$\lambda^{24} \odot t \otimes t' \otimes t'' \leftarrow t_0 \otimes t_0' \otimes t_0''. \quad (13.9)$$

It can be reduced to the λ-algorithm

$$\lambda^{24} \odot t \otimes t' \otimes t'' \leftarrow M^3 \odot (1,1,1),\ M = 2m(p+1), \tag{13.10}$$

which is the product of the λ-algorithm $t \leftarrow M \odot (1,1,1)$ and of the two dual λ-algorithms, $t' \leftarrow M \odot (1,1,1)$, $t'' \leftarrow M \odot (1,1,1)$. The associated equation of the latter algorithm defines the familiar exponent 3 log 52/log 110 for m=6, p=12. On the other hand, the λ-algorithm (13.10) can be improved and a smaller exponent can be obtained if we reduce the number of terms (1,1,1) on the right side of (13.10). To achieve that goal, note that the product

$$t_0 \otimes t_0' \otimes t_0'' = 8m^3p^3 \odot (1,1,1) \oplus 4m^3p^2 \odot ((1,1,2) \oplus$$

$$(1,2,1) \oplus (2,1,1)) \oplus 2m^3p \odot ((1,2,2) \oplus (2,1,2) \oplus (2,2,1)) \oplus m^3 \odot (2,2,2)$$

contains the term $m^3 \odot (2,2,2)$. Using the nontrivial reduction

$$m^3 \odot (2,2,2) \leftarrow 7\, m^3 \odot (1,1,1)$$

via the application of Strassen's algorithm of rank 7 for (2,2,2), we can save m^3 multiplications of 1 X 1 matrices or equivalently we can save m^3 terms (1,1,1) on the right side of (13.10). Then, solving the associated equation

$$3(2(m-1)(p-1))^{\omega} = m^3(8(p+1)^3-1),$$

we reduce the bound 3 log 52/log 110 = 2.5218127... to 2.5218006... [S81]. Although numerically that improvement is insignificant, some further progress can be made if we consider higher powers of the λ-algorithm (13.9). A certain sequence of recursive algorithms based on the λ-algorithms (13.6)-(13.8) leads to the exponents that satisfy the equation

$$27(2(m-1)(p-1))^{\omega} = (2mp + 2^{\omega-2}m)(2mp + 2m)^2 \text{ for all } m,p.$$

For m=6, p=11 this implies the bound $\omega < 2.5167$, see [R82].

A substantially longer construction, also relied on the algorithms (13.6)-(13.8), leads to the bound $\omega \leq 3\tau$ where $\tau = \tau(m,p)$ is the real root of the following equation,

$$3(2(m-1)(p-1))^{\tilde{\tau}} = 2mp + 2^{\tilde{\tau}} m, \tag{13.11}$$

see Statement A1 on page 124 of [P81]. This implies the exponent $\omega < 2.5161$ (choose $m=6$, $p=11$).

Since a shorter argument leads to even a little better exponents (see the next three sections), we will not prove (13.11) here. A detailed proof of (13.11) has been presented in the manuscript [S80] written in the spring 1980 but still remaining unpublished, apparently because a further progress was already expected in the spring 1980 and, indeed, soon followed. Namely, applying a modification of the techniques of trilinear aggregating to the construction of recursive λ-algorithms and starting with the λ-algorithm (12.6), D. Coppersmith and S. Winograd obtained the exponent 2.496. They presented that result in [CW] but announced it already in September 1980.

14. Recursive Application of Trilinear Aggregating

It is possible to suggest some promising designs that rely on aggregating the *triplets* of principal terms but our thorough investigation (which we leave outside this book) convinces that in the previous section we have already obtained the best exponent that can be derived in that way.

The improvement from 2.5160... to 2.495... comes where we aggregate more terms using a recursive construction that starts with the λ-algorithm (12.1). We will formally present that construction in Section 16. In the present section we will describe its first step in some detail. Then in the next section we will demonstrate some technique that will be used in Section 16 in order to continue the recursive construction further. If the readers wish, they may skip that demonstration and go from here directly to Section 16.

We will start with a familiar design, that is, with the bilinear λ-algorithm represented by the trilinear identity (12.1). We could define the Kronecker square of that identity and consequently the associated bilinear λ-algorithm

$$\lambda^4 \odot ((m^2,1,p^2) \oplus 2 \odot (m,mp,p) \oplus (1,m^2p^2,1)) \leftarrow (m+1)^2(p+1)^2 \odot (1,1,1), \tag{14.1}$$

which is the square of the bilinear λ-algorithm

$$\lambda^2 \odot ((m,1,p) \oplus (1,mp,1)) \leftarrow (m+1)(p+1) \odot (1,1,1) \tag{14.2}$$

associated with the identity (12.1), compare Propositions 8.1 and 10.1. We know that the λ-algorithm (mapping) (14.2) can be improved and turned into the λ-algorithm (mapping) (12.6) by imposing the appropriate equations (12.2)-(12.5) on the variables. It can be easily shown that such an improvement can be extended, so that the mapping (14.1) can be also improved and at least can be turned into the mapping

$$\left.\begin{array}{l}\lambda^4 \odot ((m^2,1,p^2) \oplus 2 \odot (m,(m-1)(p-1),p) \oplus \\ (1,(m-1)^2(p-1)^2,1)) \leftarrow (mp+1)^2 \odot (1,1,1).\end{array}\right\} \quad (14.3)$$

The associated equations of the mapping (14.3) is the square of the associated equation of the mapping (12.6), so that both mappings define the same exponent. Our objective in this section is to reduce that exponent by further improving the mapping (14.1), so that it will turn into the mapping

$$\left.\begin{array}{c}\lambda^6 \odot ((m^2,1,p^2) \oplus 2 \odot (m,(m-1)(p-1),p) \oplus (1,(m-1)^2(p-1)^2+ 2mp,1)) \\ \leftarrow (mp+1)^2 \odot (1,1,1).\end{array}\right\} \quad (14.4)$$

In the next section we will generalize this process of improving the given λ-algorithms and finally reduce the exponent below 2.5 in that way.

In order to achieve that goal, we will examine and modify the process of devising the λ-algorithms (14.1) and (14.3) by aggregating 4-tuples of principal terms.

As our first step, we will define the aggregating table for the λ-algorithm (14.1). We will derive that table from the 4 X 3 Table 14.1 below whose r-th column is the Kronecker product (see Definition 10.2) of the two r-th columns of two 2 X 3 tables for r=1,2,3. One of them is Table 12.5 and another one is Table 12.5* below, which differs from Table 12.5 only in that the asterisks are added to all latin letters.

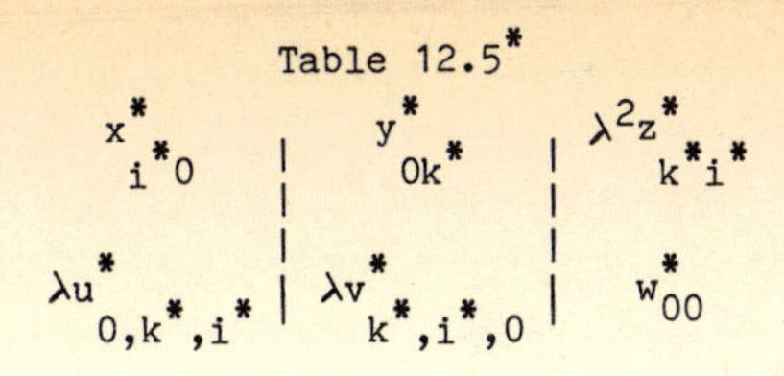

Table 12.5*

$x^*_{i^*0}$	$y^*_{0k^*}$	$\lambda^2 z^*_{k^*i^*}$
$\lambda u^*_{0,k^*,i^*}$	$\lambda v^*_{k^*,i^*,0}$	w^*_{00}

Table 14.1

$x^*_{i^*0} \otimes x_{i0}$	$y^*_{0k^*} \otimes y_{0k}$	$\lambda^4 z^*_{k^*i^*} \otimes z_{ki}$
$\lambda u^*_{0,k^*,i^*} \otimes x_{i0}$	$\lambda v^*_{k^*,i^*,0} \otimes y_{0k}$	$\lambda^2 w^*_{00} \otimes z_{ki}$
$\lambda x^*_{i^*0} \otimes u_{0,k,i}$	$\lambda y^*_{0k^*} \otimes v_{k,i,0}$	$\lambda^2 z^*_{k^*i^*} \otimes w_{00}$
$\lambda^2 u^*_{0,k^*,i^*} \otimes u_{0,k,i}$	$\lambda^2 v^*_{k^*,i^*,0} \otimes v_{k,i,0}$	$w^*_{00} \otimes w_{00}$

Both Tables 12.5* and 12.5 are aggregating tables that define the bilinear λ-algorithm (14.2). Therefore we may expect that the Kronecker product of those two tables will define an aggregating table for the bilinear λ-algorithm associated with the mapping (14.1), which is the square of (14.2).

We have to rewrite the entries of Table 14.1 as indeterminates (compare (10.9)) in order to associate the principal terms of that table with the terms of the appropriate trilinear form for Disjoint MM. This suggests transforming Table 14.1 into the following one.

Table 14.2

$x^{(0)}(i^*,i;0)$	$y^{(0)}(0;k^*,k)$	$\lambda^4 z^{(0)}(k^*,k;i^*,i)$
$\lambda x^{(1)}(i;k^*,i^*)$	$\lambda y^{(1)}(k^*,i^*;k)$	$\lambda^2 z^{(1)}(k,i)$
$\lambda x^{(2)}(i^*;k,i)$	$\lambda y^{(2)}(k,i;k^*)$	$\lambda^2 z^{(2)}(k^*,i^*)$
$\lambda^2 u(0;k^*,i^*,k,i)$	$\lambda^2 v(k^*,i^*,k,i;0)$	$w(0,0)$

Table 14.2 is, indeed, an aggregating table for the problem of Disjoint MM on the left side of (14.1). (To check that fact, assume that i^* and i range from 0 to

m-1 and that k^* and k range from 0 to p-1. Rename the variables by substituting the pairs $(i^*+im,0)$ for $(i^*,i;0)$; $(0,k^*+kp)$ for $(0;k^*,k)$; (k^*+kp,i^*+im) for $(k^*,k;i^*,i)$; (i,k^*+i^*p) for $(i;k^*,i^*)$; $(k^*+i^*p;k)$ for (k^*,i^*,k); $(i^*,k+ip)$ for $(i^*;k,i)$; $(k+ip,k^*)$ for $(k,i;k^*)$; $(0,k^*+i^*p+(k+ip)mp)$ for $(0;k^*,i^*,k,i)$ and $(k^*+i^*p+(k+ip)mp,0)$ for $(k^*,i^*,k,i;0)$. ∎

(Note the positions and the meaning of the semicolons above. We will not explicitely change the triplets and 5-tuples of indices in Table 14.2 into the above pairs of indices but we will just keep in mind that such a change is possible.)

Remark 14.1. We have introduced the special notation (u,v,w rather than $x^{(4)}$, $y^{(4)}$, $z^{(4)}$) for the variables in the last row of Table 14.2 in order to emphasize our special attention to that row, which defines the subproblem of the problem of Disjoint MM where we will obtain an improvement over the λ-algorithm (14.3).

Aggregating Table 14.2 immediately defines the λ-algorithm (14.1). (The required grouping of the correction terms is almost straightforward.)

As our next step, we will derive the λ-algorithm (14.3) from Aggregating Table 14.2. Impose the following equations on the x-variables and on the y-variables in order to cancel some of the correction terms.

$$\sum_{i^*} x^{(1)}(i;k^*,i^*) = \sum_{i} x^{(2)}(i^*;k,i) = \sum_{k^*} y^{(1)}(k^*,i^*;k) = \sum_{k} y^{(2)}(k,i;k^*)=0, \quad (14.5)$$

$$y^{(1)}(k^*,0;k) = y^{(2)}(k,0;k^*)) = x^{(1)}(i;0,i^*) = x^{(2)}(i^*;0,i) = 0 \quad (14.6)$$

for all i^*,k^*,i,k.

Also similar equations should be imposed on the u-variables and on the v-variables.

$$\sum_{i^*} u(0;k^*,i^*,k,i) = \sum_{i} u(0;k^*,i^*,k,i) = \sum_{k^*} u(0;k^*,i^*,k,i) =$$

$$\sum_{k} u(0;k^*,i^*,k,i) = \sum_{i^*} v(k^*,i^*,k,i;0) = \sum_{i} v(k^*,i^*,k,i;0) = \quad (14.7)$$

$$\sum_{k^*} v(k^*,i^*,k,i;0) = \sum_{k} v(k^*,i^*,k,i;0) = 0 \text{ for all } i^*,k^*,i,k.$$

Actually the equations (14.7) will be modified later on.

In order to reduce the number of remaining correction terms of Table 14.2, group them into the aggregates of the three following tables (which are obtained from Table 14.2 by the substitution of zeroes for some of the entries, so that all principal terms of the three new tables equal 0).

Table 14.3 (for all i,k)

$\sum_{i^*} x^{(0)}(i^*,i;0)$	$\sum_{k^*} y^{(0)}(0;k^*,k)$	0
0	0	$-\lambda^2 z^{(1)}(k,i)$
$\lambda \sum_{i^*} x^{(2)}(i^*;k,i)$	$\lambda \sum_{k^*} y^{(2)}(k,i;k^*)$	0
0	0	$-w(0,0)$

Table 14.4 (for all i^*,k^*)

$\sum_{i} x^{(0)}(i^*,i;0)$	$\sum_{k} y^{(0)}(0;k^*,k)$	0
$\lambda \sum_{i} x^{(1)}(i;k^*,i^*)$	$\lambda \sum_{k} y^{(1)}(k^*,i^*;k)$	0
0	0	$-\lambda^2 z^{(2)}(k^*,i^*)$
0	0	$-w(0,0)$

Table 14.5

$\sum_{i^*,i} x^{(0)}(i^*,i;0)$	$\sum_{k^*,k} y^{(0)}(0;k^*,k)$	0
0	0	$w(0,0)$

We will <u>increase the dimension of the problem</u> solved by the λ-algorithm but will preserve its λ-rank, compare (14.3),(14.4), if we will replace the last rows of all of the three latter tables (that is, $[0,0,w(0;0)]$ or $[0,0,-w(0;0)]$,) by the rows of the form $[\pm\lambda^2 u(0,h), \lambda^2 v(h,0), \pm w(0,0)]$ in all 2mp+1 instances of those three tables. Note that the last rows of Tables 12.5 and 14.2 also take similar forms, so we follow the approach of Section 12. (Here we assume that h is allowed to be either a parameter, or a pair, or a triplet, or a 4-tuple of parameters.)

Let us confirm our statement about the increased efficiency by some estimates. The sum of the principal terms of the modified last rows of all four Tables 14.2-14.5 is the trilinear form associated with 1XQ by QX1 MM where $Q = m^2p^2 + 2mp+1-\nu$ is the number of indeterminates among the u-variables and the v-variables, ν is the total number of linearly independent linear equations in the u-variables and in the v-variables that we have to impose on those variables, in order to cancel the correction terms of the form uyz, uyw, xvz, xvw. Let us show how we can do with a smaller ν. Note that the terms of the form uyz and xvz can be cancelled without imposing any cancelling equations on the u-variables and on the v-variables but just by changing some powers of λ in Tables 14.2-14.5. Indeed, consider the following aggregating tables obtained from Tables 14.2-14.5 by changing the powers of λ and replacing the bottom rows by "more efficient" ones as this was suggested above.

Table 14.6 (compare Table 14.2)

$x^{(0)}(i^*,i;0)$	$y^{(0)}(0;k^*,k)$	$\lambda^6 z^{(0)}(k^*,k;i^*,i)$
$\lambda x^{(1)}(i;k^*,i^*)$	$\lambda y^{(1)}(k^*,i^*;k)$	$\lambda^4 z^{(1)}(k,i)$
$\lambda x^{(2)}(i^*;k,i)$	$\lambda y^{(2)}(k,i;k^*)$	$\lambda^4 z^{(2)}(k^*,i^*)$
$\lambda^3 u(0;k^*,i^*,k,i)$	$\lambda^3 v(k^*,i^*,k,i;0)$	$w(0,0)$

Table 14.7 (compare Table 14.3)

$\sum_{i^*} x^{(0)}(i^*,i;0)$	$\sum_{k^*} y^{(0)}(0;k^*,k)$	0
0	0	$-\lambda^4 z^{(1)}(k,i)$
$\lambda \sum_{i^*} x^{(2)}(i^*;k,i)$	$\lambda \sum_{k^*} y^{(2)}(k,i;k^*)$	0
$-\lambda^3 u(0;k,i)$	$\lambda^3 v(k,i;0)$	$-w(0,0)$

Table 14.8 (compare Table 14.4)

$\sum_{i} x^{(0)}(i^*,i;0)$	$\sum_{k} y^{(0)}(0;k^*,k)$	0
$\lambda \sum_{i} x^{(1)}(i;k^*,i^*)$	$\lambda \sum_{k} y^{(1)}(k^*,i^*;i)$	0
0	0	$-\lambda^4 z^{(2)}(k^*,i^*)$
$-\lambda^3 u(0;p+k^*,m+i^*)$	$\lambda^3 v(p+k^*,m+i^*;0)$	$-w(0,0)$

Table 14.9 (compare Table 14.5)

$\sum_{i^*,i} x^{(0)}(i^*,i;0)$	$\sum_{k^*,k} y^{(0)}(0;k^*,k)$	0
$\lambda^3 u(0,0)$	$\lambda^3 v(0,0)$	$w(0,0)$

It remains to replace the equations (14.7) by new linear equations in the u-variables and in the v-variables in order to cancel the correction terms of the form $uyw(0,0)$ and $xvw(0,0)$. Consider those terms as linear forms in the x-indeterminates and in the y-indeterminates and equate all coefficients of such linear forms to 0. The factor $w(0,0)$ can be ignored in those equations, so that it is sufficient to introduce a total of $\nu = |X| + |Y|$ linear equations in the u-variables and in the v-variables. Here $|X| = m^2 + 2m(m-1)(p-1)$, $|Y| = p^2 + 2p(m-1)(p-1)$ are the numbers of all x-indeterminates and of all y-indeterminates, respectively. (To verify the above formulae for $|X|$ and $|Y|$, estimate the numbers of all <u>linearly</u> <u>independent</u> equations in the x-variables and in the y-variables among the equations (14.5) and (14.6) and subtract those two numbers from the numbers of all x-variables and all y-variables, respectively.) Therefore

$Q = m^2p^2 + 2mp + 1 - |\underline{X}| - |\underline{Y}| = (m-1)^2 (p-1)^2 + 2mp$, and we obtain the desired mapping (14.4). Applying Theorem 7.2 in the case where $m=p=3$ we yield the bound

$$\omega < 2.5199 \text{ over all rings } F. \tag{14.8}$$

This is a noticeable improvement over the exponent 2.548, see (12.7). We certainly should try to continue this recursive construction. In principle, it is possible to present the next step using the 288 X 3 aggregating tables whose three columns are the Kronecker squares of the three columns of 14 X 3 table composed of Table 14.6-14.9. This, however, would be a too tedious work. In the next section we will demonstrate a modification of the techniques of trilinear aggregating that will help us in Section 16 where we will represent the next steps of the same construction in a much more compact way.

<u>Remark</u> <u>14.2</u>. We have already mentioned at the end of Section 12 that the reduction of the exponent should have some limitation if we aggregate only a fixed number (say, 2,3,4) of principal terms. If we aggregate more and more principal terms, then we will need to use more and more correction terms. Then the total number of terms becomes too large in order to lead us to a decrease of the exponent. If we impose some equations on the variables in order to cancel some correction terms, then such equations should not be too numerous because otherwise the reduction of the dimension of the problem would again prevent the exponent from going down. The recursive construction begun in this section and continued in Section 16 seems to be a nearly optimum way of keeping delicate balance between aggregating appropriately many principal terms and canceling appropriately many correction terms by imposing equations on the variables. Note also the following special feature of the most successful constructions of Sections 12-14: the pairs, the triplets, and the 4-tuples of the principal terms (to be aggregated) represent the disjoint problems of MM of <u>varying</u> <u>sizes</u>, that is, the different lines of the same aggregating table (see Tables 12.5, 13.1, 13.2, 14.2) define the MM problems of different sizes. The aggregation of the terms of the problems of the same size in [P82b] gives asymptotically slower algorithms.

15. Can The Exponent Be Further Reduced?

In the next section we will extend the recursive construction of the previous section and reduce the exponent below 2.5. In this section we will demonstrate one of the basis steps of such an extension by revisiting and generalizing the λ-algorithm defined by Table 12.5 and by the identity (12.1). For such a demonstration we will prove the following result, which itself is of some interest (compare Theorem 2.1).

Theorem 15.1. For an arbitrary bilinear algorithm of rank M for the problem (n,n,n), $n>1$, over an arbitrary field F of constants, $\omega_F < \omega^* = \log M/\log n$. (Combining Theorem 15.1 with Strassen's algorithm implies that $\omega < \log 7$. However, Theorem 15.1 itself relies on the same construction that has already led us to the exponent 2.55.)

Proof. Represent the given bilinear algorithm in the form of the trilinear identity (10.2) where L_q, L_q' are defined by (2.2). We may always linearly transform the variables in order to satisfy the inequalities $f''(0,0,q) \neq 0$ for all q. Indeed, we may choose any linear transformation of the variables that satisfies the matrix equations $X' = XG^{-1}$, $Y' = GYH^{-1}$, $Z' = HZ$ where X', Y', Z' are the matrices of new variables, G,H are two invertible constant matrices with the entries from F or from an algebraic extension of F. Then $Tr(X'Y'Z') = Tr(XYZ)$ and we come to a bilinear algorithm of rank M for the same problem (n,n,n) represented by $Tr(X'Y'Z')$. We may choose the matrices G and H such that $f'(0,0,q) \neq 0$ for all q. Then the duality transformation of the algorithm that turns the y-variables into z-variables will also turn the latter inequalities into the desired inequalities $f''(0,0,q) \neq 0$. For simplicity we will preserve the same notation for the coefficients f' and f'' in spite of the linear and dual transformations and will assume that $f''(0,0,q) \neq 0$ for all q already in the original algorithm.

Next we will temporarily assume that $M \geq 2n^2 + n^{1/2}$ and represent the product $\lambda^2 L_q L_q' L_q''$ as the principal term of the following one-line table.

Table 15.1

L_q	L_q'	$\lambda^2 L_q''$

Consider also the following one-line table whose principal terms are the products $\lambda^2 u_{0q} v_{q0} w_{00}$. Their sum in q ranging from 0 to M-1 defines the problem $\lambda^2 \circledcirc (1,M,1)$.

Table 15.2

$(\lambda/f''(0,0,q))u_{0q}$	λv_{q0}	$f''(0,0,q)w_{00}$

Aggregate Tables 15.1 and 15.2 into the following table.

Table 15.3

L_q	L'_q	$\lambda^2 L''_q$
$(\lambda/f''(0,0,q))u_{0q}$	λv_{q0}	$f''(0,0,q)w_{00}$

Table 15.3 defines the λ-algorithm represented by the following trilinear identity,

$$\lambda^2(\mathrm{Tr}(XYZ) + \sum_{q=0}^{M-1} u_{0q}v_{q0}w_{00}) =$$

$$\sum_q (L_q + (\lambda/f''(0,0,q))u_{0q})(L'_q + \lambda v_{q0})(\lambda^2 L''_q + f''(0,0,q)w_{00}) - \tag{15.1}$$

$$\sum_q L_q L'_q f''(0,0,q)w_{00} - \lambda \sum_q L_q v_{q0} f''(0,0,q)w_{00} - \lambda \sum_q u_{0q}L'_q w_{00} \text{ modulo } \lambda^3.$$

(Recall that $\sum_q$ designates the sum in q ranging from 0 to M-1, see Definition 2.1.)

This algorithm computes the two disjoint matrix products, XY and $\sum_q u_{0q}v_{q0}$, so we expect that the associated exponent will be smaller than ω^* due to the additional output $\sum_q u_{0q}v_{q0}$. We must, however, keep the number of correction terms sufficiently small. At first note that the identity (10.2) implies that

$$\sum_q L_q L'_q f''(0,0,q)z_{00} = \sum_j x_{0j}y_{j0}z_{00}.$$

Hence

$$\sum_q L_q L'_q f''(0,0,q)w_{00} = \sum_j x_{0j}y_{j0}w_{00}.$$

Substitute the latter identity into (15.1) and note that this reduces the number of correction terms by M-n. Next note that $\lambda \Sigma_1 = \lambda \sum_q L_q v_{q0} f''(0,0,q)w_{00}$ can be rewritten as the sum of only n^2 correction terms since $L_q = \sum_q f(i,j,q)x_{ij}$,

$$\lambda \Sigma_1 = \lambda \sum_{i,j} x_{ij}(\sum_q f(i,j,q)f''(0,0,q)v_{q0})w_{00}. \tag{15.2}$$

Similarly the sum $\lambda \Sigma_2 = \lambda \sum_q u_{0q}L'_q w_{00}$ can be rewritten as the sum of n^2 terms,

$$\lambda \Sigma_2 = \lambda \sum_{j,k} (\sum_q f'(j,k,q)u_{0q})y_{jk}w_{00}. \tag{15.3}$$

Therefore the sum of all correction terms of Table 15.3 can be regrouped as the sum of a total of only $2n^2 + n$ terms. This defines a λ-algorithm of rank $M + 2n^2 + n$ and of degree 2 for the problem $\lambda^2 \otimes ((n,n,n) \oplus (1,M,1))$. The associated exponent is already less than ω^* for many pairs M and n. We will slightly improve the above λ-algorithm by imposing the following $2n^2$ linear equations on the u-variables and on the v-variables.

$$\left.\begin{aligned} &\sum_q f'(j,k,q)u_{0q} = 0 \text{ for all } j,k, \\ &\sum_q f(i,j,q)f''(0,0,q)v_{q0} = 0 \text{ for all } i,j. \end{aligned}\right\} \qquad (15.4)$$

The latter equations cancel all of the remaining correction terms but the n terms $x_{0j}y_{j0}w_{00}$. (Note that the latter n terms represent the problem (1,n,1).)

Complement the $2n^2$ equations (15.4) by $2n^2$ complementary linear equations of the form $u_{0r} = 0$, $v_{s0} = 0$ for the appropriate r,s such that under the resulting $4n^2$ linear equations the trilinear form $u_{0q}v_{q0}w_{00}$ represents the problem $(1,M-2n^2,1)$.

Summarizing obtain the following mapping,

$$\lambda^2 \otimes ((n,n,n) \oplus (1,M-2n^2,1)) \leftarrow M \otimes (1,1,1) \oplus (1,n,1).$$

This, of course, implies the mapping

$$\lambda^2 \otimes ((n,n,n) \oplus (1,M-2n^2,1)) \leftarrow (M + n) \otimes (1,1,1). \qquad (15.5)$$

Comparing the associated equation of (15.5)

$$n^{3\tau} + (M - 2n^2)^{\tau} = M + n$$

with the equation $n^{\omega^*} = M$ we obtain that $3\tau < \omega^*$ if $(M-2n^2)^{\omega^*} > n^3$. We may assume that $\omega^* > 2$ because $\rho(n,n,n) > n^2$, see (35.1) in Section 35. Therefore Theorem 15.1 follows if $M \geq 2n^2 + n^{3/2}$.

Let us show that even if the latter inequality does not hold, the similar inequality must hold for some powers of the original algorithm,

$$(n^h,n^h,n^h) \leftarrow M^h \otimes (1,1,1), \; h=1,2,\ldots$$

The associated equations of all such powers of the given algorithm define the same exponent ω^*, of course. Therefore it remains to show that $M^h > 2n^{2h} + n^{3h/2}$ for some sufficiently large h. The latter property holds because otherwise

$$M \leq \lim_{h \to \infty} (2n^{2h} + n^{3h/2})^{1/h} = n^2.$$

The latter inequality would have contradicted the known lower bound $\rho(n,n,n) \geq 2n^2 - 1$, see (35.2). ∎

Remark 15.1. Theorem 15.1 is presented in [CW] but it was cited by A. Schönhage in his lecture at MIT, Cambrdige, Mass., already in April 1980. (A. Schönhage attributed that theorem to S. Winograd.) [CW] contains also the following more general result.

Theorem 15.2. Every bilinear λ-algorithm B for MM or for Disjoint MM over a field F of constants can be transformed into another bilinear λ-algorithm B^* for Disjoint MM over F such that $\omega(B^*) < \omega(B)$ where $\omega(B)/3$, $\omega(B^*)/3$ are the unique positive roots of the two equations associated with the λ-algorithms B and B^*, respectively, provided that $\omega(B) > 2$ and that the equation associated with the λ-algorithm B has the unique positive root. (Compare Theorem 7.2 where the inequality M>S implies the uniqueness of the positive root.)

Remark 15.2. The power of Theorems 15.1 and 15.2 should not be exaggerated. Those two theorems do show that if the exact value of the exponent of MM is not equal to 2, then it may be derived from decompositions more general than bilinear λ-algorithms, compare Section 40 and the end of Section 17, or it may be derived as a limiting point of the exponents associated with some bilinear λ-algorithms but it cannot equal a root of the equation associated with a single bilinear algorithm or λ-algorithm. However, those theorems lead only to nominal numerical improvement of the upper bounds on ω except for the case where one starts with the decomposition (12.1), applies that approach recursively, and obtains a little more substantial improvement at the first two-three steps. The latter case will be studied in the next section.

16. The Exponents Below 2.5.

In this section we will reduce the exponent below 2.5 by applying an extension of the recursive construction of Section 15 to the basis algorithm of Section 12. The first step of that recursive construction will be the same as in Section 14. We will start with the following class of bilinear λ-algorithms, which generalize the bilinear λ-algorithms (12.6),(14.1)-(14.4),

$$\lambda^d \odot (t \oplus (1,H,1)) \leftarrow M \odot (1,1,1). \qquad (16.1)$$

Here t defines a problem of MM or of Disjoint MM. (These λ-algorithms may also remind us of the λ-algorithm (15.5).)

Next we will write the associated aggregating table and the associated trilinear identity of such λ-algorithms; generalizing Table 15.3 and the identity (15.1) we will allow any linear functions in the u-variables and in the v-variables over the ring $F[\lambda]$ modulo λ^{d+1} (that is, over polynomials in λ modulo λ^{d+1} over F) rather than just the expressions $(\lambda/f''(0,0,q))u_q$ and λv_q, which we used in Table 15.3 and in the identity (15.1).

Aggregating Table 16.1.

$L_q(X)$	$L'_q(Y)$	$L''_q(Z)$
$l_q(U)$	$l'_q(V)$	w_{00}

$$\lambda^d(T(X,Y,Z) + \sum_{h=0}^{H-1} u_{0h}v_{h0}w_{00}) = \qquad (16.2)$$

$$\sum_q (L_q(X) + l_q(U))(L'_q(Y) + l'_q(V))(L''_q(Z) + w_{00}) \text{ modulo } \lambda^{d+1}$$

where $T(X,Y,Z) = \sum_{q=0}^{M-1} L_q(X)L''_q(Y)L''_q(Z)$ modulo λ^{d+1} is the trilinear form associated with the problem t of (16.1) and

$$\sum_{h=0}^{H-1} u_{0h}v_{h0}w_{00} = \sum_q l_q(U)l'_q(V)w_{00} \text{ modulo } \lambda^{d+1}.$$

Here q ranges from 0 to M-1, see Definition 2.1, and $L_q(X),L'_q(Y),L''_q(Z),l_q(U),l'_q(V)$ are linear functions in the variables from the sets X,Y,Z,U,V, respectively, whose coefficients are polynomials in λ. In such a representation we consider X,Y,Z,U,V as sets rather than matrices. We assume that the set of all variables is partitioned into those five subsets and into the set $\{w_{00}\}$ that consists of the single variable w_{00}. Next we will define the following subclass of the class (16.1),(16.2) of bilinear λ-algorithms.

<u>Definition</u> <u>16.1</u>. The λ-algorithm (16.1),(16.2) is called (non-strictly) <u>improvable</u> if the linear forms $L_q(X)$ and $L'_q(Y)$ in (16.2) satisfy the following identity,

$$\sum_{q=1}^{M} L_q(X)L_q'(Y) = 0. \tag{16.3}$$

Proposition 16.1. The λ-algorithm (12.1) and the λ-algorithm (14.3) (defined by Tables 14.2-14.5 and by the equations (14.5)-(14.7)) as well as the λ-algorithm (14.4) (defined by Tables 14.6-14.9, by the equations (14.5)-(14.6), and by $|X|+|Y|$ equations in the u-variables and v-variables) are all improvable.

Proof. Both λ-algorithms (12.1) and (14.3) have the format (16.1),(16.2). The identity (16.3) is easily verified for the λ-algorithm (12.1). Indeed,

$$\sum_q L_q(X)L_q'(Y) = \sum_{i,k} x_{i0}y_{0k} - \sum_i x_{i0} \sum_k y_{0k} = 0.$$

In the sequel we will show how (16.3) can be extended from a general λ-algorithm (16.2) to its square under certain conditions that hold for the λ-algorithm (12.1). The further transition from the square of (12.1) to (14.3) and (14.4) only requires to change the powers of some λ-coefficients and to impose some linear equations on the variables, and this leaves (16.3) invariant. This proves Proposition 16.1. However, we will also complete the proof by just verifying (16.3) directly in the case of the λ-algorithms (14.3) and (14.4). Indeed, compare Tables 14.2-14.5 and derive that

$$\sum_q L_q(X)L_q'(Y) = \sum_{i^*,i,k^*,k} (x^{(0)}(i^*,i;0) + \lambda x^{(1)}(i;k^*,i^*) +$$

$$\lambda x^{(2)}(i^*;k,i))(y^{(0)}(0;k^*,k) + \lambda y^{(1)}(k^*,i^*;k) + \lambda y^{(2)}(k,i;k^*)) -$$

$$\sum_{i,k} \sum_{i^*} (x^{(0)}(i^*,i;0) + \lambda x^{(2)}(i^*;k,i)) \sum_{k^*} (y^{(0)}(0;k^*,k) + \lambda y^{(2)}(k,i;k^*))$$

$$- \sum_{i^*,k^*} \sum_{i} (x^{(0)}(i^*,i;0) + \lambda x^{(1)}(i;k^*,i^*)) \sum_{k} (y^{(0)}(0;k^*,k) + \lambda y^{(1)}(k^*,i^*;k))$$

$$+ \sum_{i^*,i} x^{(0)}(i^*,i;0) \sum_{k^*,k} y(0;k^*,k).$$

This is identically zero under the equations (14.5),(14.6), as can be easily verified. The latter expressions for $L_q(X)L_q'(Y)$ will not be changed by the transition from the λ-algorithms defined by Tables 14.2-14.5 to ones defined by Tables 14.6-14.9, as well as by any linear equations imposed on the u-variables and on the v-variables. ∎

The next result justifies the name "improvable".

Proposition 16.2. Every improvable bilinear λ-algorithm (16.1),(16.2) can be transformed into the improvable bilinear λ-algorithm,

$$\lambda^{2d+2} \odot (t \oplus (1,M-|X| - |Y|,1)) \leftarrow M \odot (1,1,1). \qquad (16.4)$$

Here t,M,d,H are defined by (16.1),(16.2); $|X|$ and $|Y|$ designate the numbers of the x-indeterminates and of the y-indeterminates in (16.2), respectively. (Therefore, if $H < M - |X| - |Y|$, then the associated exponent of the resulting λ-algorithm (16.4) is strictly less than the associated exponent of the original λ-algorithm (16.1),(16.2).)

Remark 16.1. Proposition 16.2 applied to the λ-algorithm (16.4) leaves it unchanged, so that the improvement may not be strict. We may try, however, to apply Proposition 16.2 to the powers of the λ-algorithms (16.4). Particularly we will continue the recursive construction of Section 14 that way and we will show that every step will give a strict improvement of the exponent.

Proof of Proposition 16.2. We will proceed via the transformation of (16.2) similar to the transformation of (15.1) into (15.5), compare also the transformations of (12.1) into (12.6) and (14.1) into (14.4). (Note that the λ-algorithm (15.5) can be improved whenever it is derived as in Section 15 from the algorithm (15.1) such that $L_q(X)$, $L'_q(Y)$ satisfy (16.3) because in that case the n correction terms $\sum_j x_{0j}y_{j0}w_{00}$ vanish.) During that transformation of (16.2) we will nowhere violate the identity (16.3). This fact will be easy to verify and will not require more comments. We will start the desired transformation with the following modifications of Table 16.1 and of the identity (16.2).

Table 16.2

L_q	L'_q	$\lambda^{d+2}L''_q$
$\lambda^{d+1}u_{0q}$	$\lambda^{d+1}v_{q0}$	w_{00}

$$\lambda^{2d+2}(T(X,Y,Z) + \sum_q u_{0q}v_{q0}w_{00}) =$$

$$\sum_q (L_q + \lambda^{d+1}u_{0q})(L'_q + \lambda^{d+1}v_{q0})(\lambda^{d+2}L''_q + w_{00}) - \qquad (16.5)$$

$$\sum_q L_qL'_qw_{00} - \lambda^{d+1}\sum_q L_qv_{q0}w_{00} - \lambda^{d+1}\sum_q u_{0q}L'_qw_{00} \text{ modulo } \lambda^{2d+3}.$$

Here the term $\sum_q u_{0q}v_{q0}w_{00}$ defines 1XM by MX1 matrix product.

In the particular case where (16.1) defines the λ-algorithm (14.2), such a transformation corresponds to the transition from Tables 14.2-14.5 to Tables 14.6-14.9 and to the associated λ-algorithm.

Following the definitions of Section 11, we shall call the products $\lambda^{d+2}L_qL_q'L_q''$ and $\lambda^{2d+2}u_{0q}v_{q0}w_{00}$ principal terms, $(L_q+\lambda^{d+1}u_{0q})(L_q'+\lambda^{d+1}v_{q0})(\lambda^{d+2}L_q''+w_{00})$ aggregates, and all other terms in (16.5) correction terms.

Note that the same number of aggregates is associated with each of Tables 16.1 and 16.2 and that the second row of Table 16.2 defines the term $\lambda^{2d+2}u_{0q}v_{q0}w_{00}$. The sum of the latter terms in all q defines the problem $\lambda^{2d+2} \odot (1,M,1)$. As a whole, Table 16.2 defines a λ-algorithm of rank greater than M for the problem $\lambda^{2d+2} \odot (t \oplus (1,M,1))$ whose dimension is greater than we require in Proposition 16.2. We will transform that λ-algorithm into the λ-algorithm (16.4) by imposing certain equations on the variables. This will cancel some correction terms, so that both rank and dimension will decrease. (Here we will proceed similarly to the substitution of (15.2) and (15.3) into (15.1) in the previous section and to our earlier transformations of the λ-algorithm (12.1) and of the λ-algorithm defined by Tables 14.6-14.9. Note that the two latter λ-algorithms are the particular cases of (16.5).)

Substitute $L_q = \sum_\alpha f(\alpha,q)x_\alpha$ and $L_q' = \sum_\beta f'(\beta,q)y_\beta$, $q=0,1,\dots,M-1$, into the expressions for all correction terms of (16.5) except for $-\sum_q L_qL_q'w_{00}$. Then the sum of all correction terms of (16.5) will be rewritten as follows,

$$-\sum_q L_qL_q'w_{00} - \lambda^{d+1}\sum_\alpha x_\alpha\Big(\sum_q f(\alpha,q)v_{q0}\Big)w_{00} - \lambda^{d+1}\sum_\beta\Big(\sum_q f'(\beta,q)u_{0q}\Big)y_\beta w_{00} \text{ modulo } \lambda^{2d+3}.$$

This suggests imposing the following linear equations on the variables u_{0q}, v_{q0}, compare (12.2),(12.3), and (15.4).

$$\sum_q f(\alpha,q)v_{q0} = 0 \text{ for all } \alpha,\quad \sum_q f'(\beta,q)u_{0q} = 0 \text{ for all } \beta. \tag{16.6}$$

Such equations cancel all correction terms but $-\sum_q L_qL_q'w_{00}$. The latter terms are also canceled if (16.3) holds.

Following the notation of Section 14, let $|X|$ and $|Y|$ designate the numbers of the <u>indeterminates</u> x_α and y_β in the identity (16.5), respectively. We immediately note that there are exactly $|X| + |Y|$ equations in (16.6); some of them may linearly depend on each other. Equivalently rewrite (16.6) in order to express M_1 variables

$u_{0,q(g)}, g=0,1,\ldots,M_1-1$, as linear combinations of the other $M-M_1$ variables u_{0q}. Similarly express M_2 variables $v_{q^*(h),0}$, $h=0,1,\ldots,M_2-1$, as linear combinations of the other $M-M_2$ variables v_{q0} where $M_1 \leq |Y|, M_2 \leq |X|$. Impose also the $M_1 + M_2$ complementary equations

$$v_{q(g),0} = 0, \; u_{0,q^*(h)} = 0, \; g=0,1,\ldots,M_1-1, \; h=0,1,\ldots,M_2-1.$$

At least $M-M_1-M_2 \geq M-|X|-|Y|$ variables u_{0q} and as many variables v_{q0} (where q is neither one of $q(g)$ nor one of $q^*(h)$) remain indeterminates. The trilinear form $\sum_{q=0}^{M-1} u_{0q}v_{q0}w_{00}$ will loose at most M_1+M_2 terms but still will define the problem $(1,Q,1)$ where $Q \geq M-|X|-|Y|$. This means that we have obtained the desired mapping (16.4). ∎

The next result will justify the recursive construction of the improvable λ-algorithms. We will need to prove that the square of an improvable λ-algorithm can be also represented as an improvable λ-algorithm (16.2) (up to renaming the variables and adjusting the λ-rank M and degree d, of course.) The proof is trivial, except for the verification of the identity (16.3). However, in order to assure that the identity (16.3) is preserved if we square the original λ-algorithms, we need to narrow the class of the improvable λ-algorithms as follows.

Definition 16.2. A λ-algorithm (16.2) is recursively improvable if the identity (16.3) and the two following identities hold, compare Table 16.1,

$$\sum_q L_q(X)\, l_q'(V) = \sum_q l_q(U) L_q'(Y) = 0. \tag{16.7}$$

It can be easily checked that the transformation of the λ-algorithm (16.2) into the λ-algorithm (16.4) in the proof of Proposition 16.2 nowhere violates the identities (16.7), that is, we have the following extension of Proposition 16.2.

Proposition 16.3. Every recursively improvable bilinear λ-algorithm (16.1),(16.2) can be transformed into recursively improvable bilinear λ-algorithm (16.4).

We have already verified the identity (16.3) for the λ-algorithm (12.1). Similarly both identities (16.7) can be easily verified for the λ-algorithm (12.1) and we have the following result.

Proposition 16.4. The λ-algorithm (12.1) is recursively improvable.

Now we can state the result that substantiates the recursive construction of the recursively improvable λ-algorithms.

Proposition 16.5. The product of two recursively improvable λ-algorithms over a (commutative) ring of constants F is always a recursively improvable λ-algorithm over the same ring of constants F.

Remark 16.2. Recall that we consider only commutative rings F, see Remark 2.1.

Before starting the proof of Proposition 16.5, let us examine the numerical impact of Propositions 16.3-16.5 on the reduction of the exponents of MM. Let us recursively apply Propositions 16.3 and 16.5 to the λ-algorithm (12.1) (which is recursively improvable, see Proposition 16.4) and to its powers.

Choose m=p=3 and apply Proposition 16.3 to the λ-algorithm (12.1). This defines the λ-algorithm (12.6), which is represented by the mapping

$$\lambda^2 \odot ((3,1,3) \oplus (1,4,1)) \leftarrow 10 \odot (1,1,1)$$

in the case m=p=3. The square of that mapping is the following mapping of the form (16.1),

$$\lambda^4 \odot ((9,1,9) \oplus 2 \odot (3,4,3) \oplus (1,16,1)) \leftarrow 100 \odot (1,1,1).$$

Applying Proposition 16.3, we arrive to the familiar mapping (14.4) for m=p=3,

$$\lambda^{10} \odot ((9,1,9) \oplus 2 \odot (3,4,3) \oplus (1,34,1)) \leftarrow 100 \odot (1,1,1),$$

and to the associated exponent (14.8). Squaring the latter mapping defines a new mapping in the form (16.1), that is,

$$\lambda^{20} \odot ((81,1,81) \oplus 4 \odot (9,16,9) \oplus (1,1156,1) \oplus$$

$$4 \odot (3,136,3) \oplus 4 \odot (27,4,27) \oplus 2 \odot (9,34,9)) \leftarrow 10000 \odot(1,1,1).$$

(Surely, the associated exponent will not change because of the squaring of the λ-algorithm.) The latter mapping defines a recursively improvable λ-algorithm, see Proposition 16.5. Application of Proposition 16.3 leads to improving the term (1,1156,1) to (1,3334,1) (with the increase of the degree but with no change of the λ-rank). The application of Theorem 7.2 to the resulting λ-algorithm yields the exponent 2.4998... The next recursive step along this line yields the exponent 2.4977718. We will postpone the analysis of further minor improvements being already sufficiently well motivated in order to prove Proposition 16.5.

Proof of Proposition 16.5. We will follow the line of the transition from the λ-algorithm (12.6) to the mapping (14.4) in Section 14. (The readers may reexamine that transition as an illustration to the proof.)

We will start with two recursively improvable λ-algorithms of the form (16.2). Let one of them be represented by the identity (16.2) itself and let another of them be represented by the following similar identity where the only difference from the identity (16.2) is in the asterisks added to all latin letters,

$$\lambda^{d^*}(T^*(X^*,Y^*,Z^*) + \sum_{h^*=0}^{H^*-1} u^*_{0h^*} v^*_{h^*0} w^*_{00}) = \qquad (16.8)$$

$$\sum_{q^*=1}^{M^*-1} (L^*_{q^*}(X^*) + l^*_{q^*}(U^*))\,(L^{*'}_{q^*}(Y^*) + l^{*'}_{q^*}(V^*))\,(L^{*''}_{q^*}(Z^*) + w^*_{00}))\text{ modulo }\lambda^{d^*+1}.$$

The latter identity is defined by the following table.

Aggregating Table 16.3, compare Table 16.1.

$L^*_{q^*}(X^*)$	$L^{*'}_{q^*}(Y^*)$	$L^{*''}_{q^*}(Z^*)$
$l^*_{q^*}(U^*)$	$l^{*'}_{q^*}(V^*)$	w^*_{00}

We assume that both identities (16.2) and (16.8) define recursively improvable λ-algorithms, that is, that the identities (16.3),(16.7) hold, as well as the following identities,

$$\sum_{q^*} L^*_{q^*}(X^*)L^{*'}_{q^*}(Y^*) = \sum_{q^*} L^*_{q^*}(X^*)l^{*'}_{q^*}(V^*) = \sum_{q^*} l^*_{q^*}(U^*)L^{*'}_{q^*}(Y^*) = 0. \qquad (16.9)$$

It is possible to define the product of the λ-algorithms (16.2) and (16.8) using Proposition 10.1. However, for us it will suffice just to examine Table 16.4 below associated with that product. Column r of that 4 X 3 table is the Kronecker product of the two columns r of Tables 16.1 and 16.3, r=1,2,3 (compare Definition 10.2 and Table 14.1).

Table 16.4

$L^*_{q^*}(X^*) \otimes L_q(X)$	$L^{*'}_{q^*}(Y^*) \otimes L'_q(Y)$	$L^{*''}_{q^*}(Z^*) \otimes L''_q(Z)$
$l^*_{q^*}(U^*) \otimes L_q(X)$	$l^{*'}_{q^*}(V^*) \otimes L'_q(Y)$	$w^*_{00} \otimes L''_q(Z)$
$L^*_{q^*}(X^*) \otimes l_q(U)$	$L^{*'}_{q^*}(Y^*) \otimes l'_q(V)$	$L^{*''}_{q^*}(Z^*) \otimes w_{00}$
$l^*_{q^*}(U^*) \otimes l_q(U)$	$l^{*'}_{q^*}(V^*) \otimes l'_q(V)$	$w^*_{00} \otimes w_{00}$

Each entry of Table 16.4 can be equally considered either (i) a bilinear form or (ii) a linear form, both having the same coefficients, compare Remark 10.1 and the equations (10.9) or compare our comments that follow Definition 10.2. At first we will assume that (ii) the entries are linear forms in the variables x,y,z,u,v,w. Here x,y,z are the variables of the first three rows and u,v,w are the variables of the fourth row of the table. We may reduce Table 16.4 to the format of Table 16.1 as follows.

Table 16.5.

$L_{q^*,q}$	$L'_{q^*,q}$	$L''_{q^*,q}$
$l_{q^*,q}$	$l'_{q^*,q}$	w

Here we assume that

$$\left.\begin{aligned}
L_{q^*,q} &= L^*_{q^*}(X^*) \otimes L_q(X) + l^*_{q^*}(U^*) \otimes L_q(X) + L^*_{q^*}(X^*) \otimes l_q(U),\\
L'_{q^*,q} &= L^{*\prime}_{q^*}(Y^*) \otimes L'_q(Y) + l^{*\prime}_{q^*}(V^*) \otimes L'_q(Y) + L^{*\prime}_{q^*}(Y^*) \otimes l'_q(V),\\
L''_{q^*,q} &= L^{*\prime\prime}_{q^*}(Z^*) \otimes L''_q(Z) + w^*_{00} \otimes L''_q(Z) + L^{*\prime\prime}_{q^*}(Z^*) \otimes w_{00},\\
l_{q^*,q} = l^*_{q^*}(U^*) \otimes l_q(U),\ & l'_{q^*,q} = l^{*\prime}_{q^*}(V^*) \otimes l'_q(V),\ w = w^*_{00} \otimes w_{00}.
\end{aligned}\right\} \quad (16.10)$$

The identities (16.3),(16.7) for the product of the λ-algorithms (16.2) and (16.8) are now represented as follows.

$$\sum_{q^*,q} L_{q^*,q} L'_{q^*,q} = \sum_{q^*,q} L_{q^*,q} l'_{q^*,q} = \sum_{q^*,q} l_{q^*,q} L'_{q^*,q} = 0. \quad (16.11)$$

It remains to prove the three latter identities. This amounts to proving certain equations in the coefficients of the entries of Table 16.4. The coefficients remain the same under both assumptions that (i) the entries of Table 16.4 are bilinear forms or (ii) they are linear forms. Thus we will shift (until the very end of the proof) to assumption (i), which now becomes more convenient for us. Under that assumption, we may substitute the conventional products for all Kronecker's products in (16.10), deleting the symbols $\otimes$, see our comments after Definitions 10.1 or 10.2. Then the identities (16.11) immediately follow if we expand the resulting bilinear expressions for $L_{q^*,q}$, $L'_{q^*,q}$, $l_{q^*,q}$, and $l'_{q^*,q}$ and substitute them into (16.11), then sum all terms separately in q^* and in q, and finally apply the identities (16.3), (16.7), and (16.9). Let us show all of those steps by proving one of the three identities of (16.11) (the verification of the two other identities of (16.11) is similar and will be omitted). Substitute the bilinear expressions for $L_{q^*,q}$ and $l'_{q^*,q}$ from (16.10) into (16.11) and expand the products. (Here we will delete the symbols ⊗ and will use the commutativity of the ring of constants, compare Remark 16.2.)

$$\sum_{q^*,q} L_{q^*,q} l'_{q^*,q} = \sum_{q^*,q} (L^*_{q^*}(X^*)L_q(X) + l^*_{q^*}(U^*)L_q(X) + L^*_{q^*}(X^*)l_q(U))l^{*\prime}_{q^*}(V^*)l'_q(V)$$

$$= \sum_{q^*,q} (L^*_{q^*}(X^*)L_q(X)l^{*\prime}_{q^*}(V^*)l'_q(V) +$$

$$l^*_{q^*}(U^*)L_q(X)l^{*\prime}_{q^*}(V^*)l'_q(V) + L^*_{q^*}(X^*)l_q(U)l^{*\prime}_{q^*}(V^*)l'_q(V))$$

$$= \sum_{q^*} L^*_{q^*}(X^*)l^{*\prime}_{q^*}(V^*)\sum_q L_q(X)l'_q(V) + \sum_{q^*} l^*_{q^*}(U)l^{*\prime}_{q^*}(V^*)\sum_q L_q(X)l'_q(V) +$$

$$\sum_{q^*} L^*_{q^*}(X^*)l^{*\prime}_{q^*}(V^*)\sum_q l_q(U)l'_q(V).$$

The latter 4-linear form is the sum of the three products of the pairs of bilinear forms. In each of the three pairs at least one is identically zero because $\sum_q L_q(X)l'_q(V) = 0$ (see (16.7)) and $\sum_{q^*} L^*_{q^*}(X^*)l^{*\prime}_{q^*}(V^*) = 0$ (see (16.9)). Hence $\sum_{q^*,q} L_{q^*,q} l'_{q^*,q}$ is identically zero. ∎

17. How Much Can We Reduce the Exponent?

How much can we reduce the exponent by continuing the recursive construction based on the λ-algorithm (12.1)? It is easy to see that at each subsequent recursive step the exponent is also strictly reduced. Indeed, squaring the current recursively improvable λ-algorithm B we obtain a new recursively improvable λ-algorithm $B^{\otimes 2}$ that defines the same exponent as the previous λ-algorithm does. It can be verified that the inequality $H < M - |X| - |Y|$ of Propositions 16.2 and 16.3 holds for the square $B^{\otimes 2}$ at all recursive steps. Therefore the subsequent application of these propositions to $B^{\otimes 2}$ will <u>strictly reduce</u> the exponent.

However, <u>the numerical value of the exponent more or less significantly decreases only at the first two or three recursive steps</u>, compare Remark 15.2.

A tiny further progress can be achieved if we modify the λ-algorithm (14.4) for m=p=3, that is,

$$\lambda^6 \odot ((9,1,9) \circledast 2 \odot (3,4,3) \circledast (1,34,1)) \leftarrow 100 \odot (1,1,1).$$

We note that we may identify the variables $y^{(1)}(k^*,i^*;k)$ and $y^{(2)}(k,i;k^*)$ in Tables 14.2-14.9 where the triplets (k^*,i^*,k) coincide with the triplets (k,i,k^*). This way we will transform the latter recursively improvable λ-algorithm into the following recursively improvable one,

$$\lambda^6 \odot ((9,1,9) \circledast (6,4,3) \circledast (1,34,1)) \leftarrow 100 \odot (1,1,1).$$

The latter λ-algorithm will remain recursively improvable, since our transformation surely preserves the identities (16.3) and (16.7).

Note that the total number of the y-variables has been substantially reduced. Apply Proposition 16.3 and transform the latter recursively improvable λ-algorithm into the following one,

$$\lambda^{14} \odot ((9,1,9) \circledast (6,4,3) \circledast (1,46,1)) \leftarrow 100 \odot (1,1,1).$$

The associated equation of the latter λ-algorithm defines the exponent 2.5104... Since the λ-algorithm remains recursively improvable, we may apply the recursive construction of the previous section and, after the next application of Propositions 16.3 and 16.5, we will arrive at the λ-algorithm

$$\lambda^{58} \odot ((81,1,81) \circledast (36,16,9) \circledast (1,3502,1) \circledast 2 \odot (6,184,3) \circledast$$

$$2 \odot (9,46,9) \circledast 2 \odot (54,4,27)) \leftarrow 10000 \odot (1,1,1).$$

This implies the exponent 2.4994. We may modify the latter λ-algorithm by replacing the terms $2 \odot (6,184,3)$ by $(6,184,6)$ (if we identify some of the x-variables pairwise with each other and therefore substantially reduce their total number). Then the application of Proposition 16.3 yields the following recursively improvable λ-algorithm,

$$\lambda^{58} \odot ((1,4606,1) \circledast (36,16,9) \circledast (81,1,81) \circledast (6,184,6) \circledast$$

$$2 \odot (9,46,9) \circledast 2 \odot (54,4,27)) \leftarrow 10000 \odot (1,1,1).$$

The latter λ-algorithm immediately defines the exponent 2.4979 and then, after one more application of Propositiosn 16.3 and 16.5, the exponent is reduced to 2.4967. Some further similar tricks involving also duality Theorem 2.3 for λ-algorithms finally lead to the bound 2.496, see [CW] for the detailed presentation of the later improvement.

Theorem 17.1, compare Remark 16.2. For arbitrary ring of constants F, the exponent $\omega = \omega_F$ of MM over F is less than 2.496.

Although the construction can be varied in many similar ways, the progress goes out of power at this point.

It looks like we have approached to the limits of the power of the techniques of trilinear aggregating applied to the design of λ-algorithms for Disjoint MM, compare Remarks 14.2 and 15.2. Thus we have arrived at the point where we should end our attempts of a further asymptotic acceleration of MM unless new ideas have been worked out. The previous history of the study of MM suggests that if such ideas appear at all, then they are most likely to appear in the form of new original algorithmic designs, such as ones from [St69] (Strassen's for 2 X 2 MM), from [P72], see (4.1),(4.2), or from [BCLR], see Example 4.1. (Recall that the latter two designs appeared after 10 years with no progress and then it took less than 2 years, from October 1978 to September 1980, in order to understand the techniques introduced by those designs, to utilize those techniques in a nearly optimum way, and to complete all work for their substantiation, including Theorems 6.3 and 7.2.)

The latter experience suggests searching for new successful algorithms for some problems of MM of small dimensions where one may attack the problem by trying to solve the system (5.5) in f, f', f'' for relatively small M and n. Similar equations can be derived for the coefficients of λ-algorithms for MM and Disjoint MM. The problem, however, becomes quite hard even where the dimension of the MM problem is small, see [P81d]. Another direction of the attack at the problem will be indicated at the end of Section 40.

Those readers who would be interested in even a minor reduction of the exponent may try to improve the recursive construction of the previous section using Theorem 17.2 below. We will apply the following definition in the statement of that theorem.

Definition 17.1. The problem of the multiplication of two matrices that may be partially filled with zeroes and all of its dual problems are called the problems of Partial MM. Equivalently, the bilinear problem represented by the trilinear form $Tr(XYZ) = \sum_{i,j,k} x_{ij}y_{jk}z_{ki}$ where either some of the x-variables and some of the y-variables, or some of the y-variables and some of the z-variables, or some of the z-variables and some of the x-variables can be replaced by zeroes is called a problem of Partial MM. The total number of nonzero terms of Tr(XYZ) in its straightforward representation is called the cardinality of the problem of Partial MM. (The definition of Partial MM is due to [S79]; the first nontrivial algorithm for Partial MM was presented earlier, see Example 4.1 that we took from [BCLR].)

Theorem 17.2, see [S79,S81]. Given a bilinear λ-algorithm of λ-rank M and of degree d for Partial MM of cardinality N, then

$$\omega \leq 3 \log M/\log N.$$

Theorem 17.2 generalizes Theorems 2.1 and 6.3 because the problem (m,n,p) can be considered a problem of Partial MM of cardinality mnp. Disjoint MM can be also considered a special case of Partial MM. For instance, the problem $2 \odot (n,n,n)$ is the problem of Partial MM obtained from the trilinear form $Tr(XYZ) = \sum_{i,j,k=0}^{2n-1} x_{ij}y_{jk}z_{ki}$ by replacing zeroes for z_{ki} whenever $|k-i| \geq n$. Therefore it is possible to apply Theorem 17.2 to the case of Disjoint MM but Theorem 7.2 gives stronger bounds on ω in that case. (For instance, application of Theorem 17.1 to an algorithm of rank M for $2 \odot (n,n,n)$ yields only the bound $\omega \leq 3 \log M/\log (2n^3)$ while application of Theorem 7.2 to the same algorithm yields the bound $\omega \leq \log(M/2)/\log n$. On the other hand, Theorems 7.2 and 17.2 can be combined together in order to define the exponents in the case where we have a bilinear λ-algorithm for the evaluation of several disjoint products of partially filled matrices. Such a combination of the two theorems might help to utilize some modification of the recursive construction of the previous section in order to reduce the exponent of MM (probably not more than, say, by 0.0001).

It is conceivable that the further progress will follow due to the advances in the algebraic theory of tensors and trilinear forms. Note, however, that we have to deal with the very specific trilinear forms such as $\sum_s Tr(X(s)Y(s)Z(s))$ where it is hardly possible to apply the estimates for the ranks of general 3-dimensional tensors, compare [St82].

We will conclude with demonstrating a generalization of the idea of trilinear aggregating which might help to accelerate MM further. At first consider the following generalization of Aggregating Tables 11.3 and 11.4.

Aggregating Table 17.1

$f(0)x(i(0),j(0))$	$f'(0)y(j(0),k(0))$	$f''(0)z(k(0),i(0))$
$f(1)u(i(1),j(1))$	$f'(1)v(j(1),k(1))$	$f''(1)w(k(1),i(1))$

Here $f(s)$, $f'(s)$, $f''(s)$ for $s=0,1$ are constants that belong to an algebra F whose elements commute with the variables x,y,z,u,v,w. We need to turn all correction terms into 0 but to leave the products $f(i)f'(i)f''(i)$ nonzero for $i=0,1$. Such a requirement is satisfied, for instance, if F is the algebra of 2X2 real matrices and

$$f(0) = f'(0) = f''(0) = \begin{bmatrix} 1 & 0 \\ 0 & 0 \end{bmatrix}, \quad f(1) = \begin{bmatrix} 0 & 1 \\ 0 & 0 \end{bmatrix} \quad f'(1) = \begin{bmatrix} 0 & 0 \\ 0 & 1 \end{bmatrix}, \quad f''(1) = \begin{bmatrix} 0 & 0 \\ 1 & 0 \end{bmatrix}.$$

Then the sum of the two principal terms of Table 17.1 $x(i(0),j(0))y(j(0),k(0))z(k(0),i(0)) + u(i(1),j(1))v(j(1),k(1))w(k(1),i(1))$ can be read in the left upper entry of the 2X2 matrix that represents the aggregate of Table 17.1, $LL'L''$, where $L = f(0)x(i(0),j(0))+f(1)u(i(1),j(1))$, $L' = f'(0)y(j(0),k(0))+f'(1)v(j(1),k(1))$, $L'' = f''(0)z(k(0),i(0))+f''(1)w(k(1),i(1))$. The multiplication of the 3 latter factors is too costly in this example but the approach might work more efficiently for larger aggregating tables that consist of S rows of the following form (where s ranges from 0 to S-1),

$$f(s)x^{(s)}(i(s),j(s)) \mid f'(s)y^{(s)}(j(s),k(s)) \mid f''(s)z^{(s)}(k(s),i(s)).$$

According to the general principle of trilinear aggregating, the coefficients $f(s),f'(s),f''(s)$ of that table should be chosen (in an algebra F whose elements commute with the variables x,y, and z) such that

i) several correction terms have been canceled (the more of them the better) and

ii) for all s the products $f(s)f'(s)f''(s)$ are nonzero so that the principal terms $x^{(s)}(i(s),j(s))y^{(s)}(j(s),k(s))z^{(s)}(k(s),i(s))$ can be reconstructed from the aggregates of the table.

For instance, such requirements have been satisfied in the case of Aggregating Table 17.1 by our choice of the coefficients f,f',f''. In general we need that

$$f(i(q),j(q))f'(j(r),k(r))f''(k(s),i(s))=0 \qquad (17.1)$$

for several triplets q,r,s where $q \neq r$, or $r \neq s$, or $s \neq q$ but we do not require that (17.1) hold necessarily for all such q,r,s so that some correction terms are supposed to be canceled by other means, compare [P82b]. We leave further investigation along this line as a challenge for the reader who may start with the attempts to satisfy (17.1) for sufficiently many triplets q,r,s in the case where F is an algebra of KXK matrices such that K^{ω} is substantially less than S.

PART 2.

CORRELATION BETWEEN MATRIX MULTIPILCATION AND OTHER COMPUTATIONAL PROBLEMS.

BIT-TIME, BIT-SPACE, STABILITY, and CONDITION

Summary. In this part of the book we will extend our asymptotically fast algorithms for MM to the algorithms for fast computation for some related problems. We will also estimate the amount of storage-space, the numbers of bit-operations involved in our algorithms, and the condition of those algorithms that characterizes their stability. In particular in Section 18 (whose results are isolated in this book) we will consider some combinatorial computational problems, that is, Boolean matrix multiplication (hereafter referred to as Boolean MM) and the all pair shortest distance problem for a digraph (hereafter referred to as the APSD-problem). In Sections 19-21 and 26-30 we will consider the solution of a system of linear equations, matrix inversion, evaluation of the determinant of a matrix (hereafter referred to as SLE or SLE(n), MI or MI(n), Det or Det(n), respectively; n indicates the dimension of the nXn input matrix) and some other computational problems of linear algebra. We will reduce the solution of all of those combinatorial and algebraic problems to MM, defining the algorithms for those problems that involve $O(n^{\omega+\varepsilon})$ arithmetical operations where ω is the exponent of MM and ε is arbitrary positive number, see Sections 18-21. Also we will estimate the bit-time of those algorithms, see Sections 18, 26-30, and of MM itself, see Sections 23 and 24. (The bit-time is defined as the number of bit-operations required for the approximate evaluation of the solution with a prescribed precision E in a given domain D; bit-time depends on E and D.) Estimating the bit-complexity is sometimes quite tedious but, rather unexpectedly, we finally show that the arithmetical complexity classification of the above problems differs from their bit-complexity classification. Specifically, for the customary choices of E and D the bit-time of our algorithms for Boolean MM, SLE, and MI remains at worst of the same order as in the case of MM. Furthermore, see Section 30, SLE can be solved faster than MM (and, in fact, in a near optimum way) in terms of the bit-time involved in the cases where E sufficiently rapidly converges to 0. On the other hand, our algorithm that reduces the APSD-problem for a digraph with n-vertices to MM(n) involves at least n times more bit-operations than it is required for nXn MM and so does also any algorithm for Det(n). In Section 22 we will show that fast MM is compatible with the minimization of the storage space involved. (That result can be extended to the case of the related computational problems.) In Section 25 we will study the condition of our algebraic algorithms and will see some correlation between the condition and the quoitient: bit-time divided by the number of arithmetical operations.

18. Reduction of Some Combinatorial Computational Problems to MM.

In this section we will show that Boolean matrix multiplication (Boolean MM) can be reduced to MM, see [AHU], pp. 242-243. (The reduction is due to [FM] and implies the reduction of several combinatorial computational problems to MM.) We will also comment on the reduction to MM of the all pair shortest distance problem of a digraph (the APSD-problem). (The reduction is due to [Y].) We will not use those reductions in the sequel.

The Boolean product $W = [w_{ik}]$ of two nXn Boolean matrices $X = [x_{ij}]$ and $Y = [y_{ik}]$ with the entries 0 and 1 is defined by the usual formulae $w_{ik} = \sum_j x_{ij}y_{jk}$ for all i and k but under the assumption that the arithmetical operations are performed in the closed semiring $\{0,1\}$ using the following rules, $0+0=0$, $1+0=0+1=1+1=1$, $0*0=0*1=1*0=0$, $1*1=1$, compare [AHU], pp. 195-196. In that case the evaluation of W is computationally equivalent to the evaluation of the transitive closure of a directed graph (see [AHU], pp. 202-204). On the other hand, let $U = [u_{ik}]$ be the usual matrix product of X and Y. Then, as is easily verified, we have for all i,k that

$$\text{either } w_{ik} = u_{ik} = 0, \text{ or } w_{ik} = 1,\ 1 \le u_{ik} \le n. \tag{18.1}$$

Therefore, the evaluation of W has been reduced to the evaluation of U, which can be performed by asymptotically fast methods for MM. The time-complexity of Boolean MM should be measured by the number of bit-operations rather than by the number of arithmetical operations involved. For the evaluation of U, however, we will use only additions, subtractions, and multiplications of integers, performed modulo $n+1$. Then the number of bit-operations involved will be $bt(\lceil \log(n+1) \rceil)$ times the number of arithmetical operations involved. Here the factor $bt(\lceil \log(n+1) \rceil)$ equals the number of bit-operations required in order to add, subtract, or multiply two integers modulo $n+1$. Here and hereafter (as before) all logarithms are to the base 2 and we did use and will use the customary notation $\lceil u \rceil$. We will also use the customary symbols $\lfloor u \rfloor$, O, o.

Notation 18.1. For a real u, $\lfloor u \rfloor$ and $\lceil u \rceil$ denote the two integers such that $\lfloor u \rfloor \le u \le \lceil u \rceil$ and $\lceil u \rceil - \lfloor u \rfloor$ is minimum; for any pair of functions $f(s)$ and $g(s)$, the notation $f(s)=O(g(s))$ as $s \to \infty$ (or equivalently $f(s)$ is $O(g(s))$ as $s \to \infty$) means that $|f(s)| \le c|g(s)|$ for a constant c and for all sufficiently large s; the notation $f(s)=o(g(s))$ as $s \to \infty$ means that $\lim_{s \to \infty} f(s)/g(s) = 0$. (Similarly as $s \to \infty$ or $s \to a$ for an arbitrary a.)

We arrive at the following theorem.

Theorem 18.1, [FM]. The minimum number $bt(BBM(n))$ of bit-operations required for nXn Boolean MM is $O(ar_Z(MM(n)))bt(\lceil \log(n+1) \rceil)$ as $n \to \infty$. Here and hereafter $ar_F(MM(n))$ designates the minimum number of arithmetical operations required for $MM(n)$ over a ring F of constants, Z is the ring of integers, and

$$bt(s) = \text{maximum } \{bt(\pm,s), bt(*,s)\} \tag{18.2}$$

where $bt(\pm,s)$ and $bt(*,s)$ designate the numbers of bit-operations required in order to perform an addition (or a subtraction) and a multiplication of a pair of integers modulo 2^s, respectively.

The next bound,

$$bt(s) = O(s \log s \log \log s) \text{ as } s \to \infty, \tag{18.3}$$

follows from the two theorems below. In all estimates we will assume that the arithmetical operations are performed on Turing machines, compare Remark 18.1 below.

Theorem 18.2, see [Of]. An addition (a subtraction) of two integers modulo 2^s can be performed involving $bt(\pm,s) = O(s)$ bit-operations as $s \to \infty$.

Theorem 18.3, see [SS], [AHU],[BM],[K],[T]. Two integers can be multiplied modulo 2^s involving $bt(*,s)=O(s \log s \log \log s)$ bit-operations as $s \to \infty$.

Remark 18.1, see [S80],[K]. We may reduce the latter upper bounds if we allow to use computers that are more powerful than Turing machines. $O(s \log s)$ bit-operations will suffice if random access machines are allowed and $O(s)$ bit-operations will suffice if storage modification machines (also called pointer machines) are allowed.

The next theorem shows that divisions are not much harder than multiplications.

Theorem 18.4. (S. Cook, see [K], pp. 295-296). Let u and v be two positive integers such that $u < 2^s$. Then the unique pair of nonnegative integers s (quotient) and r (remainder) such that $u=qv+r$ and $r<v$ can be evaluated involving $bt(/,s)=O(bt(*,s))$ bit-operations as $s \to \infty$, provided that $d(*,s)$ bit-operations suffice in order to multiply two arbitrary integers modulo 2^s.

In the sequel we will also refer to the following result from [Br76].

Theorem 18.5. Let $bt(sqr,s)$ designate the number of bit-operations required for the evaluation of $v^{1/2}$ where $-2^s < v < 2^s$ and both v and $v^{1/2}$ are represented as normalized floating point binary numbers whose fractions have been chopped to s binary bits. Then $bt(sqr,s) = O(bt(*,s))$ as $s \to \infty$.

Remark 18.2. Theorems 18.2-18.5 can be complemented by the simple observation that, say, at least s bit-operations are necessary on any of the above machines in order to perform any arithmetical operation over an arbitrary pair of integers modulo 2^s because all 2s bits of the input information must be processed, so that $bt(s) \geq s$.

Coming back to estimating the asymptotic complexity of Boolean MM (in terms of the numbers of bit-operations involved) we note that the factor $bt(\lceil \log(n+1) \rceil)$ is negligible and obtain the following corollary from Theorem 18.1.

Theorem 18.6. $\omega(BBM) \leq \omega_Z$. Here ω_Z is the exponent of MM over the ring Z of integer constants and $\omega(BBM)$ is the exponent of Boolean MM, that is,

$$\omega(BBM) = \lim_{n \to \infty} (\log bt(BMM(n))/\log n). \tag{18.4}$$

Applying Theorem 23.6 from Section 23, see Remark 23.1, we may slightly strengthen Theorem 18.6 as follows.

Theorem 18.7. $\omega(BBM) \leq \omega_Q$ where ω_Q is the exponent of MM over the field of rational constants.

It is tempting to extend Theorems 18.6 and 18.7 from Boolean MM to the more general APSD-problem of finding the shortest distances d(i,k) from i to k for all pairs (i,k) of vertices of a given digraph G. Here d(i,k) is the minimum total cost of going from i to k by a path consisting of the edges of G provided that the costs of all edges of G are some given natural numbers c(i,k), that $c(i,k) = +\infty$ if there is no edge in G from i to k, and that $d(i,k) = \infty$ if there is no path in G from i to k, see [AHU], pp. 195-209; [La], pp.59-108; [Ta], pp. 85-96.

For a digraph G with n vertices, the APSD-problem can be easily reduced to the solution of the following system of equations,

$$\left.\begin{aligned} & d^{(0)}(i,i) = 0,\ d^{(0)}(i,k) = +\infty,\ i \neq k, \\ & d^{(m+1)}(i,k) = \underset{0 \leq j \leq n-1}{\text{minimum}}(d^{(m)}(i,j) + c(j,k)). \end{aligned}\right\} \tag{18.5}$$

Here i,k=0,1,...,n-1; m=0,1,...,N-1, and any natural value exceeding n-1 can be chosen for N, see [La], p. 82. More precisely, $d(i,k) = d^{(N)}(i,k)$ for all i,k if $N \geq n$, so it is sufficient to evaluate the matrix $D^{(N)} = [d^{(N)}(i,k)]$ for some $N \geq n$. Recursively apply (18.5) and represent

$$D^{(N)} = D^{(N-1)} \boxtimes C = D^{(N-2)} \boxtimes C \boxtimes C = \ldots = D^{(0)} \boxtimes C \boxtimes C \boxtimes \ldots \boxtimes C = U^{(0)} \boxtimes C^{\boxtimes N}.$$

Here $D^{(m)}$ and C are the nXn matrices, $D^{(m)} = [d_{ik}^{(m)}]$, $C = [c_{kj}]$; the symbol $\boxtimes$ designates the generalized matrix multiplication such that $D^{(m+1)} = D^{(m)} \boxtimes C$ is defined by (18.5) for m=0,1,...,N-1, compare [La], p. 83. Then it suffices to evaluate first the generalized power $C^* = C^{\boxtimes N}$ of C and then the generalized product $D^{(0)} \boxtimes C^*$. We may choose $N = 2^s$ where $\log n \leq s \leq 1 + \log n$, evaluate $C^{\boxtimes N}$ by the generalized squaring of C repeated s times involving at most 1+log n generalized nXn MM's, and, after one more generalized nXn MM, solve the APSD-problem for a digraph with n vertices. On the other hand, the generalized nXn MM can be reduced to the conventional nXn MM so that we have the following result.

Theorem 18.8, [Y]. For arbitrary positive ε and arbitrary digraph G with n vertices and with given natural costs c(i,k) of edges from vertex i to vertex k for i,k=0,1,...,n-1, it suffices to perform $O(n^{\omega+\varepsilon})$ arithmetical operations, $2n^2 \lceil 2 + \log n \rceil$ evaluations of the powers 2^{-u} for some natural numbers u, and

$n^2 \lceil 2 + \log n \rceil$ evaluations of the logarithms of finite binary numbers (all of those operations are assumed to be performed with the infinite precision) in order to evaluate the matrix $D=[d(i,k)]$ of the shortest distances from the vertex i to the vertex k of the digraph G for all n^2 pairs (i,k). Here $\omega = \omega_Q$ is the exponent of MM over the field of rational constants.

Remark 18.3. Theorem 18.8 can be applied also in the case where some costs c(i,k) are negative but where there is no cycles of negative cost in the digraph G. If such cycles exist, then the computed matrix $D^{(N)}$ for $N \geq n$ will not equal the matrix D of the shortest distances but in that case some of the diagonal entries of the matrix $D^{(N)}$ will be negative (otherwise they always equal 0). Substituting $-\infty$ for the values 0 of the diagonal entries of C at the places where the diagonal entries of $D^{(N)}$ are negative, we may then repeat the same process and finally evaluate the desired matrix D. On the other hand, Theorem 18.8 can be extended to the case where the given entries of the cost matrix C are finite binary numbers (not necessarily integers). If the entries of C are real, then an approximation to the matrix D can be computed by the given method if we represent the entries of C as (generally infinite) binary numbers and chop their fractions to a prescribed number of binary bits.

Remark 18.4. It is somewhat artificial to measure the complexity of the APSD-problem by the number of infinite precision operations because such an operation amounts to the infinite number of bit-operations while the input information to the APSD-problem can be represented by the matrix C defined by only n^2c binary bits where $c = 1 + \text{maximum} \lceil \log c(i,k) \rceil$ and the maximization is in all finite entries c(i,k) of the matrix C. In Theorem 18.10 below we will estimate the bit-operation complexity of the APSD-problem. To illustrate the possibility of saving some arithmetical operations at the expense of the increase of the number of bit-operations per arithmetical operation, we will reproduce the algorithm from Examples 4.2 and 4.3 of [P80a], pp. 12-13, for the multiplication of two polynomials $P_{m-1}(\lambda)$ and $Q_{n-1}(\lambda)$, compare Example 2.2 in Section 2.

Examples 18.1. Let $m \geq n$, let at first the coefficients x_α and y_β of the two given polynomials be natural, and let a natural k exceed $\log(m\, x_\alpha y_\beta)$ for all α and β. Then the coefficients of the polynomials $P_{m-1}(\lambda)$, $Q_{n-1}(\lambda)$, and $P_{m-1}(\lambda)Q_{n-1}(\lambda)$ can be read from the binary representation of the natural numbers $P_{m-1}(2^k)$, $Q_{n-1}(2^k)$, and $P_{m-1}(2^k)Q_{n-1}(2^k)$, so that one multiplication suffices in order to solve the problem. The point is that the multiplication has been performed here over very large factors, that is, over km-bit and kn-bit integers, with no saving in the total number of bit-operations as the result. Still this approach seems efficient in the cases where the multiplication of long integers can be performed on a computer that easily operates with numbers that are substantially longer than the products $mx_\alpha y_\beta$, compare the end of Remark 18.6 below. Note also that the case where the input coefficients are integers can be reduced to the case where they are natural by representing $P_{m-1}(\lambda)$ and $Q_{n-1}(\lambda)$ as the differences of two pairs of polynomials with natural coefficients. Furthermore, if the coefficients are rational, they can be turned into integers by scaling. If they are real or complex numbers, they can be approximated by finite binary numbers.

Remark 18.5. Due to the simplicity of the latter example, it had to appear somewhere earlier than in 1980 but we have no reference at hand and refer to the report [P80a] where that algorithm was presented and cited in the summary.

Proof of Theorem 18.8. It is sufficient to reduce the generalized MM to MM. We will apply the following auxiliary result.

Proposition 18.9 [R80a]. Let m and n be natural numbers, such that $m > \log n$; $c(j)$ for each natural j is either $+\infty$ or a natural number; $d = \sum_{j=0}^{n-1} 2^{-c(j)m}$; $c = \min_{j=0,1,\ldots,n-1} c(j)$. Then $c = \lceil -(\log d)/m \rceil$.

Proof. It follows from the definition of c and d that $2^{-cm} \le d \le n2^{-cm}$. Hence $c-(\log n)/m \le -(\log d)/m \le c$. ∎

Now assume that we are given two nXn matrices $U=[u(i,j)]$ and $V=[v(j,k)]$ whose entries are natural or $+\infty$. Then the generalized product $W = U \otimes V = [w(i,k)] = [\min_j (u(i,j)+v(j,k))]$ (compare (18.5)) can be evaluated as follows.

Algorithm 18.1.

Step 1. Choose $m = 1 + \lfloor \log n \rfloor$. Evaluate the two nXn matrices $X=[x(i,j)]$ and $Y=[y(j,k)]$ such that

$$\left.\begin{array}{l} x(i,j)=0 \text{ if } u(i,j)=+\infty,\ y(j,k)=0 \text{ if } v(j,k)=+\infty, \\ -m\,u(i,j) = \log x(i,j),\ -m\,v(j,k) = \log y(j,k) \text{ for all } i,j,k. \end{array}\right\} \quad (18.6)$$

Step 2. Evaluate the matrix $Z=[z(i,k)]=XY$.

Step 3. Evaluate the matrix $W=[w(i,k)]$ as follows,

$$w(i,k) = \lceil -(\log z(i,k))/m \rceil,\quad i,k=0,1,\ldots,n-1. \quad (18.7)$$

The matrix W is the desired generalized product $U \otimes V$ as this follows from (18.6),(18.7), and Proposition 18.9. Therefore the APSD-problem for a digraph G with n vertices can be solved in $\lceil 2 + \log n \rceil$ applications of Algorithm 18.1. Count the number of operations at Steps 1-3 and derive Theorem 18.8. ∎

Keeping our promise given in Remark 18.4, we will next estimate the bit-operation complexity of the APSD-problem. We will modify Algorithm 18.1, see [P81a], computing approximations to the sums $\sum_j x(i,j)y(j,k)$ with the relative error at most 0.5 (such a precision will be assured by the relations (18.8)-(18.10) below) rather than the exact values of these sums. Such approximations will serve us not worse than the exact values if m is increased just by 1.

Algorithm 18.2. Proceed as in Algorithm 18.1 but (i) choose

$$m = 2 + \lfloor \log n \rfloor \tag{18.8}$$

and (ii) at Step 2 evaluate a matrix $Z=[z(i,k)]$ where $z(i,k)$ for all i and k approximate to $\sum_j x(i,j)y(j,k)$ with the absolute errors

$$\Delta(i,k) = z(i,k) - \sum_j x(i,j)y(j,k) \tag{18.9}$$

such that

$$\log|\Delta(i,k)| < -mr(i,k) - 1, r(i,k) = \underset{j}{\text{minimum}}\ (u(i,j) + v(j,k)). \blacksquare \tag{18.10}$$

(Note that we do not need to evaluate $r(i,k)$ for all i,k but only need know an upper bound on $r(i,k)$.) The relations (18.6)-(18.10) and Proposition 18.8 imply that

$$\underset{j}{\text{minimum}}(u(i,j) + v(j,k)) = \lceil -(\log z(i,k))/m \rceil = w(i,k) \text{ for all } i,k,$$

so that Algorithm 18.2 indeed evaluates the desired matrix W. ■

Until the end of this section, let

$$u = \underset{i,j}{\text{maximum}}\ u(i,j),\ v = \underset{j,k}{\text{maximum}}\ v(j,k),\ r = \underset{i,k}{\text{maximum}}\ r(i,k) \tag{18.11}$$

where the maximums are over all finite $u(i,j)$, $v(j,k)$, and $r(i,k)$. (18.10) will hold if

$$\log|\Delta(i,k)| < - mr - 1 \text{ for all } i,k. \tag{18.12}$$

Next note that the evaluation of 2^q for an integer q (see Step 1) amounts just to the representation of that power of 2 as a normalized floating point binary number with the exponent q+1, so that Step 1 involves at most $\sum_j(\sum_i \lceil \log u(i,j) \rceil + \sum_k \lceil \log v(j,k) \rceil) = n^2\, O(\log u + \log v)$ bit-operations. Furthermore, at Step 3 we need not evaluate logarithms because (18.7) implies that $w(i,k) = \lceil -e(i,k)/m \rceil$ where $e(i,k)$ is the exponent of the number $z(i,k)$ represented with the floating point as a normalized binary number. The divisions by m are relatively simple, see Theorem 18.4, so that the dominating part of the total number of bit-operations comes from Step 2.

Step 2 amounts to performing nXn MM with the absolute precision at most 2^{-mr-1}, see (18.12), provided that the entries of the input matrices lie between 0 and 1, r is defined by (18.11), and m is defined by (18.8). As follows from Theorem 23.6, see Section 23, the bit-complexity of that step is $O(n^{\omega+\varepsilon}\,bt(r))$ as $n \to \infty$(for arbitrary positive ε). Here $\omega = \omega_Q$ is the exponent of MM over the field of rational constants, bt(s) is defined by (18.2),(18.3).

Reexamine all generalized MM's that need be performed for solving the APSD-problem for a given digraph G and note that (for all such MM's) r is bounded by the maximum finite value d in all d(i,k), that is, by the maximum in all finite shortest distances between the vertices of G. Summarizing we have the following estimate, see [P81a].

Theorem 18.10. For arbitrary positive ε, for a digraph G with n vertices, and for a matrix C of the costs of the edges of G, which are either natural numbers or are equal to $+\infty$, the matrix D of the shortest (directed) distances between all pairs of vertices of G can be evaluated involving $O(n^{\omega+\varepsilon}\,bt(d))$ bit-operations where $\omega = \omega_Q$ is the exponent of MM over the field of rational constants, d is the maximum in all finite entries of the matrix D, and bt(s) is defined by (18.2),(18.3).

Remark 18.6. In the general case, d can be as large as n, even if all entries of the cost matrix are either 0, or 1, or $+\infty$. If d has the order of magnitude n, then $n^{\omega+\varepsilon}d > n^3$ while the bit-operation complexity of the APSD-problem is known to be $O(n^3(\log\log n/\log n)^{1/3})$, see [Fr]. Thus so far the value of Theorem 18.9 has been limited either to the cases where the shortest paths are required to be bounded, say, by a fixed constant or to the models of analysis where the multiple precision arithmetic is considered inexpensive. The latter models could be justified in the future development of computer technology, compare [DGK].

One may ask if fast Boolean MM or any fast method for the APSD-problem may help in speeding up MM. This may seem unlikely because MM requires to process continuous information while the two other problems deal with the discrete one. Such an argument may not be valid, however, if we consider only the multiplication of matrices whose each entry equals either 0 or 1, compare (18.1). If such a problem were reduced to Boolean MM or to the APSD-problem, then this could lead to efficient algorithms for MM in the average case because Boolean MM and the APSD-problem can be solved fast in the average case, see [Bl] (in that case the comparison of the models of averaging for MM and for Boolean MM or for the APSD-problem will be required). On the other hand, any algorithm for the evaluation of the products of matrices with 0 and 1 entries can be extended to the evaluation of the products XY of two matrices X and Y with integer entries bounded by $1-2^s$ and 2^s-1 because we may represent such products XY as follows,

$$XY = \sum_{q=0}^{s-1} \alpha(q)2^q X(q) \sum_{r=0}^{s-1} \beta(r)2^r Y(r) = \sum_{q,r=0}^{s-1} \alpha(q)\beta(r)2^{q+r}X(q)Y(r).$$

Here $\alpha(q)$ and $\beta(r)$ equal 1 or -1 for every q and r and X(q) and Y(r) are two matrices with 0 and 1 entries for all q and r. The crucial step would be the reduction of MM to Boolean MM or to the APSD-problem, which so far remains problematic with little hopes for the progress.

19. Asymptotic Arthmetical Complexity of Some Computations in Linear Algebra.

In this section we will estimate the number of arithmetical operations required for the solution of the following problems of linear algebra: matrix inversion, solution of a system of linear equations, evaluation of the determinant of a matrix, and evaluation of all coefficients of the characteristic polynomial of a matrix. Hereafter we will refer to those problems as MI, SLE, Det, and CCP, respectively, or, in the cases where the problems are defined by nXn matrices (and, for SLE(n), also by n-dimensional vectors), as MI(n), SLE(n), Det(n), and CCP(n), respectively. Similarly MM(m,n,p) and MM(n) will designate mXn by nXp MM and nXn MM, respectively. We will extend the concept of the exponent from MM to MI, SLE, Det, and CCP, so that the exponent

$$\omega_F(\text{Problem }(n)) = \lim_{n \to \infty} (ar_F(\text{Problem }(n))/\log n). \tag{19.1}$$

Here Problem (n) may stand for any of the problems MM(n), MI(n), SLE(n), Det(n), or CCP(n) and ar_F(Problem(n)) stands for the minimum number of arithmetical operations required for the solution of that problem over a field of constants F, compare (18.4). The following result demonstrates the close correlation among the listed problems.

Theorem 19.1. For an arbitrary field of constant F,

$$\omega_F(\text{SLE}) \leq \omega_F(\text{MI}) = \omega_F(\text{MM}) = \omega_F(\text{Det}) = \omega_F(\text{CCP}).$$

For simplicity we will omit the subscript F in the remainder of this section where the results hold for any field F. We will cite the two following recent bounds without presenting their proofs,

$$ar(\text{MI}(n)) \leq 5\, ar(\text{Det}(n)) \text{ for all } n, \text{ see [BS]}, \tag{19.2}$$

$$ar(\text{CCP}(n)) = O(ar(\text{MM}(n))\log n) \text{ as } n \to \infty, \text{ see [Ke]}. \tag{19.3}$$

The next two inequalities,

$$ar(\text{Det}(n)) \leq ar(\text{CCP}(n)),\ ar(\text{SLE}(2n)) \leq ar(\text{MI}(n)) + 2n^2 - n \text{ for all } n, \tag{19.4}$$

immediately follow if we recall that the determinant of a matrix X is just the λ-free coefficient of the characteristic polynomial $\det(X - \lambda I)$ and that $\underline{v} = X^{-1}\underline{y}$ is the solution to the linear system $X\underline{v} = \underline{y}$.

Let us derive the remaining inequalities needed in order to prove Theorem 19.1. Consider the following block-factorization of matrices X and X^{-1}, see [AHU],[BM],[K],[St69]. Here X is a $(2n)\times(2n)$ matrix to be inverted; $X_{11}, X_{12}, X_{21}, X_{22}$ are $n\times n$ submatrices of X; 0 and I are the $n\times n$ null and the $n\times n$ identity matrices, respectively. (Similar notation 0 and I and sometimes 0_n and I_n will be used for the null and the identity matrices in the sequel.)

$$X = \begin{bmatrix} X_{11} & X_{12} \\ X_{21} & X_{22} \end{bmatrix} = \begin{bmatrix} I & 0 \\ X_{21}X_{11}^{-1} & I \end{bmatrix} \begin{bmatrix} X_{11} & 0 \\ 0 & Z \end{bmatrix} \begin{bmatrix} I & X_{11}^{-1}X_{12} \\ 0 & I \end{bmatrix}, \tag{19.5}$$

$$X^{-1} = \begin{bmatrix} I & -X_{11}^{-1}X_{12} \\ 0 & I \end{bmatrix} \begin{bmatrix} X_{11}^{-1} & 0 \\ 0 & Z^{-1} \end{bmatrix} \begin{bmatrix} I & 0 \\ -X_{21}X_{11}^{-1} & I \end{bmatrix} \tag{19.6}$$

where

$$Z = X_{22} - X_{21}X_{11}^{-1}X_{12}. \tag{19.7}$$

The matrix equations (19.6) and (19.7) imply the following formula,

$$\mathrm{ar}(\mathrm{MI}(2n)) \le 2\mathrm{ar}(\mathrm{MI}(n)) + 6\mathrm{ar}(\mathrm{MM}(n)) + 2n^2. \tag{19.8}$$

Apply (19.8) recursively in n and derive that

$$\mathrm{ar}(\mathrm{MI}(n)) = O(\mathrm{MM}(n)) \text{ as } n \to \infty. \tag{19.9}$$

On the other hand, (19.5) implies that $\det X = \det X_{11} \det Z$.

Taking into account also (19.7) we obtain the recursive estimates

$$\mathrm{ar}(\mathrm{Det}(2n)) \le 2\,\mathrm{ar}(\mathrm{Det}(n)) + 2\mathrm{ar}(\mathrm{MM}(n)) + \mathrm{ar}(\mathrm{MI}(n)) + n^2+1, \tag{19.10}$$

which imply that

$$\mathrm{ar}(\mathrm{Det}(n)) = O(\mathrm{MM}(n)) \text{ as } n \to \infty. \tag{19.11}$$

Theorem 19.1 immediately follows from (19.2)-(19.4),(19.9), (19.11) and from the next inequality,

$$\omega(MM) \leq \omega(MI). \qquad (19.12)$$

Let us derive (19.12) by reducing MM to MI. Apply the above matrix equations (19.6), (19.7) to the (2n)X(2n) matrix X where $X_{11}=I$. Note that in that case the lower right nXn submatrix of X^{-1} equals $Z^{-1}=(X_{22}-X_{21}X_{12})^{-1}$. Thus we have reduced the evaluation of the product $X_{21}X_{12}$ of two arbitrary nXn matrices X_{21} and X_{12} to the inversion of the (2n)X(2n) matrix X and of the nXn matrix Z^{-1}, which is a submatrix of X^{-1}, and to the subtraction of the resulting nXn matrix Z from another (given) nXn matrix X_{22}. Therefore $ar(MM(n)) \leq ar(MI(2n))+ar(MI(n))+n^2$ for all n. This implies both (19.12) and, as a consequence, Theorem 19.1. ∎

Actually the relations (19.2)-(19.4),(19.9),(19.11),(19.12) combined are stronger than Theorem 19.1. Furthermore our proof of (19.12) implies the relation

$$ar(MM(n)) = O(MI(n)) \text{ as } n \to \infty,$$

which is stronger than (19.12). Note, however, that that proof is applied to the classes of algorithms where only divisions by the constant 0 are not allowed but the divisions by rational functions are allowed even if the divisors may turn into zeroes for certain values of the input variables. For the class of such algorithms, the equation (19.3) can be also strengthened as follows,

$$ar(CCP(n)) = O(ar(MM(n)), \text{ see [Ke]}.$$

Summarizing, we have the following extension of Theorem 19.1.

<u>Theorem</u> <u>19.2</u>. $ar(Problem(n)) = O(ar(Problem_1(n)))$, $ar(SLE(n)) = O(ar(Problem(n)))$ as $n \to \infty$ provided that Problem(n) and $Problem_1(n)$ stand for an arbitrary pair among the four problems MM(n), MI(n), Det(n), and CCP(n) and that for the solution of any of the problems it is allowed to use any algorithms where divisions by all polynomials and by all rational functions (except for the divisions by the constant 0) are permitted.

In the remainder of this section we will assume that the input matrices and vectors can have arbitrary complex entries and we will show how to modify our algorithms of this section so that they will involve divisions only by nonvanishing divisors. We will see that the equations of Theorem 19.1 hold over the algorithms of that class.

At first let us show how to assure that the divisors never vanish in the fast recursive algorithms for MI and SLE that rely on (19.5)-(19.7) provided, of course, that the given matrix X is invertible. It is sufficient to show that the submatrices X_{11} and Z of X are invertible for all values of the input variables, and that the similar property holds at all recursive steps where we invert matrices. Such properties are known to be guaranteed, see [Wil], p. 29, if the given matrix X is a Hermitian positive definite matrix, that is, if $X = Y^H Y$ for a nonsingular square matrix Y. (Here and hereafter W^H designates the <u>conjugate transpose of a matrix</u> $W = [w_{jk}]$, such that W^H has the complex conjugate of w_{jk} as its (k,j) entry for all j and k. Hereafter, to be consistent, let z^H = Re z-i Im z designate the <u>complex conjugate of a complex number</u> z = Re z + i Im z.) Thus we will avoid divisions by zero in our algorithms for MI and SLE if we invert the matrix $X^H X$ and then evaluate $X^{-1} = (X^H X)^{-1} X$. A slightly more involved but even more efficient elimination of all zero divisors can be obtained using algorithms with branching and appropriate permutations of the rows and/or columns of the given matrix X, see [AHU], pp. 232-242; [BM], p. 50. In Section 21 we will show another reduction of MI and SLE to MM that never involves divisions by vanishing values provided that X is invertible. The two latter approaches also enable us to compute det X without involving zero divisors even if the matrix X is singular. (Also, if the matrix X is Hermitian and positive definite, then we may recursively use the equation det X = det X_{11}det Z in order to evaluate $\det(X^H X) = (\det X)(\det X)^H$, see (19.5)-(19.7); no divisions by zero will be involved in such an algorithm; |det X| can be obtained as the square root from the latter product.)

Finally, we will modify our proof of (19.12) in order to assure that no divisions by zero will be introduced when we reduce MM to MI. Let U and V be two given nXn matrices to be multiplied and let A(MI(n)) and A(MI(2n)) be two given algorithms for the inversion of nXn and (2n)X(2n) matrices that involve ar(MI(n)) and ar(MI(2n)) arithmetical operations, respectively. Choose four constant matrices G,H,J,K of the size nXn such that the nXn matrix Z=K-(U+H)(V+J) and the (2n)X(2n) matrix $X = \begin{bmatrix} G & G(V+J) \\ U+H & K \end{bmatrix}$ are invertible. At first evaluate X^{-1} by the algorithm A(MI(2n)) and then the inverse Z=K-(U+H)(V+J) of the nXn lower right (southwest) submatrix Z^{-1} of X^{-1} by the algorithm A(MI(n)). (The fact that such a lower right submatrix is indeed $Z^{-1} = (K-(U+H)(V+J))^{-1}$ can be easily verified if we expand (19.6) and then compare (19.7) in the case of our matrix X.) Finally evaluate K-Z-HV-UJ-HJ using the straightforward algorithms for MM and for matrix subtraction. This gives us the desired matrix product UV. The latter algorithm may involve too many arithmetical operations but note that at most ar(MI(2n)) + ar(MI(n)) of them can be nonlinear. Now we may apply the well-known reduction of any algorithm for MM(n) to a bilinear one, which increases the number of nonlinear operations at most twice, see Propositions 32.3, 34.1, and 34.2 in Sections 32 and 34. Therefore we arrive at a bilinear algorithm for MM(n) that involves at most 2ar(MI(n))+2ar(MI(2n)) bilinear steps. Application of Theorem 2.1 yields the following inequality,

$$\log(ar(MM(N)))/\log N \le (\log(2ar(MI(n)) + 2ar(MI(2n)))/\log n + c/\log N$$

for a constant c and for all n and N. Let $n \to \infty$ and $N \to \infty$, recall (19.1), and derive (19.12). ■

20. Two Block-Matrix Algorithms for the QR-factorization and for the QR-type-factorization of a Matrix.

In this section we will study the QR-factorization of an mXn matrix, see [A], p. 525-527; [DB], p. 201; [Ric], pp. 364-368; [GvL], pp. 136-188; [Wil], pp. 47-50. In Sections 21, 27-30, we will apply the results of that study in order to reduce MI, SLE, and Det to MM. That reduction is better conditioned than the reduction via the block-factorization (19.5) and will be prefered when we estimate the bit-complexity of MI, SLE, and Det (see Sections 27-30). We will use only the QR-factorization of square matrices but we will consider the case of arbitrary m and n, $m \ge n$, for the sake of completeness and for methodological convenience.

The unscaled QR-factors of an mXn matrix X are defined as an upper triangular nXn matrix R and an mXn matrix Q such that

$$X = QR, \quad Q^H Q = \Sigma \tag{20.1}$$

where Σ is an nXn diagonal matrix. Hereafter the problem of the ealuation of such unscaled QR-factors of an mXn matrix X will be designated $QR^*(m,n)$. The same problem will be designated QR(m,n) if it is required in addition that all diagonal entries of R be nonnegative and Σ be the identity matrix. (If $Q^H Q = I$, then Q is called a unitary matrix.) In some cases we will write simply QR^* and QR rather than QR(m,n) and $QR^*(m,n)$. Also we will write $QR^*(n)$ and QR(n) if m=n.

The complexity of the problems QR^* and QR is characterized by the three following theorems, compare [S73].

Theorem 20.1. Let X be an mXn matrix, $m \ge n$, and $n \to \infty$. Then (m/n)O(ar(MM(n))) arithmetical operations suffice in order to find some unscaled QR-factors of X.

Theorem 20.2. If some unscaled QR-factors of an mXn matrix X have been found, then O(mn) additional arithmetical operations as $n \to \infty$ and at most n additional evaluations of square roots suffice in order to find the scaled QR-factors of X.

Theorem 20.3. The problem QR(n) can be solved involving at most n^2 evaluations of square roots and O(ar(MM(n))) arithmetical operations as $n \to \infty$.

(Actually even n evaluations of square roots and O(ar(MM(n))) arithmetics are sufficient by the virtue of Theorems 20.1 and 20.2 but, as we will see in Remark 20.2, Algorithm 20.1 to be used in the proof of Theorem 20.1 is not always stable so that in the sequel we will refer to the stable Algorithm 20.2, which solves that problem involving n^2 evaluations of square roots.)

Theorems 20.1 and 20.2 imply an extension of our list of problems reducible to MM. The next theorem will add one more problem to that list (which could be expanded further, compare [AHU],[BM],[IMH]) and will have applications to our study of SLE(n) and Det(n), see Sections 21 and 28.

Theorem 20.4. Let X be an nXn matrix and h be an integer such that $n \leq 2^h$. Then O(ar(MM(n))) additions, subtractions, and multiplications suffice in order to find an upper triangular nXn matrix R and 4^h nonsingular 2X2 matrices Q_k, $k=1,2,\ldots,4^h$, such that $Q_k^H Q_k$ are diagonal matrices for all k and

$$Q^H X = R,\ Q^H = \tilde{Q}_{4^h}\tilde{Q}_{4^h-1}\cdots\tilde{Q}_1. \tag{20.2}$$

Here $\tilde{Q}_k$ are nonsingular nXn matrices such that

$$\tilde{Q}_k = \begin{bmatrix} I_{g(k)} & 0 & 0 \\ 0 & Q_k & 0 \\ 0 & 0 & I_{n-2-g(k)} \end{bmatrix},\ 0 \leq g(k) \leq n-2,\ k=1,2,\ldots,4^h.$$

(Note that the matrix $Q^H Q = I_n$ if $Q_k^H Q_k = I_2$ for all k but that otherwise $Q^H Q$ may not even be diagonal.)

We will prove Theorems 20.1,20.2, and 20.4 by applying two slightly generalized algorithms from [S73] to the case where m and n are powers of 2. (The algorithms of [S73] are the block-matrix representations of the Gram-Schmidt and the Givens methods of the QR-factorization.)

Remark 20.1. The latter case can be immediately extended to the general case by padding the matrix X with zeroes as follows,

$$\begin{bmatrix} X & 0 \\ 0 & I \\ 0 & 0 \end{bmatrix},\ \text{or}\ \begin{bmatrix} X & 0 \\ 0 & I \end{bmatrix},\ \text{or}\ \begin{bmatrix} X & 0 \\ 0 & 0 \end{bmatrix},$$

and by solving the same problems for such larger matrices. (Similar tricks can be applied to other computational problems of linear algebra that we considered in the previous section.)

Here is the first algorithm, which factors the mXn matrices that have the full rank n provided that $m \geq n$. That algorithm is just the block-matrix version of the Gram-Schmidt orthogonalization process, see [DB], pp. 201-202; [GvL], pp. 150-152; [Ric], pp. 365-366. The algorithm leads to Theorems 20.1 and 20.2, see also Remark 20.1 below. (Actually Theorem 20.2 is easily verified. It is also easy to check that an evaluation of a square root is required already for the scaled QR-factorization of a 1X1 matrix.) To understand the construction of the algorithm and its close relation to the Gram-Schmidt orthogonalization process, the reader may substitute the matrix equations (20.6) below into the matrix equation X=QR and then derive the formulae (20.3)-(20.5) below (which define the algorithm) from the resulting matrix equation and from the three matrix equations $Q^HQ = \Sigma$, $Q^HX = \Sigma R$, see (20.1), and $X = [X_1,X_2]$.

Algorithm 20.1. If n=1, then choose Q=X, $R = I_1$, that is, R is the 1X1 matrix with the entry 1 and Q=X is a column-vector. (If the scaled QR-factors are sought, then evaluate $R = (X^HX)^{1/2}$ and $Q=(1/R)X$ for n=1. In that case the (n/2)X(n/2) identity matrix $I_{n/2}$ should replace the matrices Σ_1 and Σ_2 below.) If n>1, then represent the matrix X as $X = [X_1,X_2]$ where both X_1 and X_2 are mX(n/2) matrices. (X_1 and X_2 must be of the full rank n/2 because X is of the full rank n.) Apply Algorithm 20.1 to the matrix X_1 and evaluate its unscaled QR-factors, Q_1 and R_{11}, and an (n/2)X(n/2) diagonal matrix, Σ_1, such that

$$X_1 = Q_1R_{11},\ Q_1^HQ_1 = \Sigma_1. \tag{20.3}$$

Then evaluate the two matrices

$$R_{12} = (\Sigma_1^{-1})Q_1^HX_2,\ \tilde{X}_2 = X_2 - Q_1R_{12}. \tag{20.4}$$

Finally apply Algorithm 20.1 to the matrix $\tilde{X}_2$ and define its unscaled QR-factors, Q_2 and R_{22}, and an (n/2)X(n/2) diagonal matrix, Σ_2, such that

$$\tilde{X}_2 = Q_2R_{22},\ Q_2^HQ_2 = \Sigma_2. \tag{20.5}$$

Then it is easy to verify that the matrix $\tilde{X}_2$ is of the full rank n/2 and that the following matrix equations define the QR-factors of X,

$$Q = [Q_1,Q_2],\ R = \begin{bmatrix} R_{11} & R_{12} \\ 0 & R_{22} \end{bmatrix}\ . \ \blacksquare \tag{20.6}$$

Remark 20.2. Algorithm 20.1 is immediately extended to the cases where the rank of the given matrix is less than n. It is sufficient just to add the following instruction in the case n=1. "If X is the null vector" (this will not occur if the given matrix is of the full rank n), "then choose R to be the 1X1 null matrix and choose $Q=\underline{u}(s)$ where $\underline{u}(s)$ is the m-dimensional unit column-vector with the s-th entry equal 1 and other entries equal 0. Here it is assumed that, after the recursive applications of (20.6), $Q=\underline{u}(s)$ ought to be placed as the s-th column of the output matrix Q." Compare, however, the two 1X1 matrices R=0 and R=I defined this way for X=0 and for X having arbitrarily small norms, respectively. Similarly the two matrices Q (Q-factors) differ nearly by $\underline{u}(s)$, so that both output matrices Q and R jump where the rank of X jumps from n down to n-q, q>0, even though the input matrix X changes as little as we like.

Let us estimate the complexity of Algorithm 20.1 for the unscaled factorization. (Similarly for the scaled factorization; just include the evaluations of the square roots and scale X as was suggested at the steps where n=1 in the algorithm above.)

Notation 20.1. Hereafter Sc(p,q), MA(p,q), and MS(p,q) will designate the multiplication of a pXq matrix by a diagonal matrix, the addition, and the subtraction of a pair of pXq matrices, respectively.

For even n, the equations (20.3)-(20.6) imply that some unscaled QR-factors of a given mXn matrix will be found if the following computational problems are solved: two problems $QR^*(m,n/2)$ and each of the problems MM(n/2,m,n/2), MM(n,n/2,n/2), Sc(n/2,n/2), and MS(m,n/2). Extending the notation of Section 8, compare (8.1), and using Notation 20.1, we represent the latter implication as follows.

$$QR^*(m,n) \leftarrow 2 \odot QR^*(m,n/2) \circledast MM(n/2,m,n/2) \circledast MM(m,n/2,n/2) \circledast MS(m,n/2) \circledast Sc(n/2,n/2).$$

We may reduce each rectangular MM to some square MM's (and MA's) and obtain the following mapping (implication),

$$\left.\begin{array}{c} QR^*(m,n) \leftarrow 2 \odot QR^*(m,n/2) \circledast (4m/n) \odot MM(n/2) \\ \circledast MS(m,n/2) \circledast (2m/n) \odot MA(n/2,n/2) \circledast Sc(n/2,n/2). \end{array}\right\} \quad (20.7)$$

(20.7) defines the bounds on $ar(QR^*(m,n))$ that, if applied recursively, yield Theorem 20.1. ∎

Next we will present the second algorithm from [S73], which defines the factorization (20.2) and the scaled QR-factorization of an arbitrary nXn matrix X with no jump of the solution even where the rank of the matrix X is reduced. The algorithm implies Theorems 20.3 and 20.4 and represents just a block-matrix rearrangement of the successive Givens rotations. The rotations are defined by the premultiplications by the unitary rotation matrices $\tilde{Q}_s$, such that $Q^H = \tilde{Q}_p\tilde{Q}_{p-1}\cdots\tilde{Q}_1$, $R = \tilde{Q}_p\tilde{Q}_{p-1}\cdots\tilde{Q}_1X$, see [GvL], p. 156, and compare [A], pp. 531,532,558; [Wil], pp. 47-48. We may choose unscaled matrices $\tilde{Q}_s$ such that $Q_s^HQ_s$ are diagonal matrices but may not be the identity matrices. Q^HQ may not be even diagonal in that case, compare (20.2).

We will present the algorithm in the case where m=n, see [S73] for the extension to arbitary m and n. As we arranged earlier, we will assume that $n=2^k$ for an integer k.

It is methodologically convenient to present the algorithm as one for the solution of the following more general problem (and then simplify it to derive the QR-factorizations and (20.2)).

<u>Problem 20.1</u>. <u>Input</u>: an $n\times n$ matrix X and an $n\times n$ upper triangular matrix $\tilde{R}$. <u>Outputs</u>: $(2n)\times(2n)$ unitary matrix $\tilde{Q}$ and $n\times n$ upper triangular matrix R such that

$$\tilde{Q}\begin{bmatrix} X \\ \tilde{R} \end{bmatrix} = \begin{bmatrix} R \\ 0 \end{bmatrix}, \tilde{Q}\,\tilde{Q}^H = I_{2n}. \tag{20.8}$$

Hereafter we will also refer to Problem 20.1 as to the problem $\tilde{Q}R(n)$. If $\tilde{R}$ is the $n\times n$ null matrix 0_n, then the unitary matrix $\tilde{Q}$ takes the form

$$\tilde{Q}^H = \begin{bmatrix} Q & 0_n \\ 0_n & I_n \end{bmatrix}.$$

Then (20.8) immediately implies (20.1), that is, the scaled QR-factors of an arbitrary $n\times n$ matrix X can be read from the solution to Problem 20.1 (to the problem $\tilde{Q}R(n)$) where $\tilde{R}=0_n$).

Here is the promised algorithm for Problem 20.1. It evaluates the scaled QR-factors of an arbitrary $n\times n$ matrix X if applied to X and $\tilde{R}=0_n$. If, in addition the formula (20.9) below is modified (actually is simplified) by substituting $R=1$ for $R=(X^HX+\tilde{R}^H\tilde{R})^{1/2}$, then the algorithm still evaluates the unscaled factors of (20.2).

<u>Algorithm 20.2 for Problem 20.1</u>, $\tilde{Q}R(n)$. For $n=1$, either let $R=0_n$ and $\tilde{Q}=I_{2n}$ if $X = \tilde{R}=0_n$ or, otherwise, evaluate

$$R=(X^HX+\tilde{R}^H\tilde{R})^{1/2}, \tilde{Q}=(1/R)\begin{bmatrix} X^H & \tilde{R}^H \\ -\tilde{R} & X \end{bmatrix}. \tag{20.9}$$

(Here the matrices R, $\tilde{R}$, and X have the size 1×1 and actually turn into numbers. In Section 26 we will use the notation r, $\tilde{r}$, and x, respectively, when we refer to (20.9).) If $n>1$, then apply Algorithm 20.2 four times for the half-size problems according to the following scheme where X and $\tilde{R}$ are represented as 2×2 block-matrices,

$$X = \begin{bmatrix} X_{11} & X_{12} \\ X_{21} & X_{22} \end{bmatrix}, \ \tilde{R} = \begin{bmatrix} \tilde{R}_{11} & \tilde{R}_{12} \\ 0 & \tilde{R}_{22} \end{bmatrix}.$$

$$\begin{bmatrix} X_{11} & X_{12} \\ X_{21} & X_{22} \\ \tilde{R}_{11} & \tilde{R}_{12} \\ 0 & \tilde{R}_{22} \end{bmatrix} \rightarrow \begin{bmatrix} X_{11} & X_{12} \\ \tilde{R}_1 & X_0 \\ 0 & X_1 \\ 0 & \tilde{R}_{22} \end{bmatrix} \rightarrow \begin{bmatrix} R_{11} & R_{12} \\ 0 & X_2 \\ 0 & X_1 \\ 0 & \tilde{R}_{22} \end{bmatrix} \rightarrow \begin{bmatrix} R_{11} & R_{12} \\ 0 & X_2 \\ 0 & \tilde{R}_2 \\ 0 & 0 \end{bmatrix} \rightarrow \begin{bmatrix} R_{11} & R_{12} \\ 0 & R_{22} \\ 0 & 0 \\ 0 & 0 \end{bmatrix}.$$

Here Algorithm 20.2 is successively applied to the nX(n/2) block-submatrices,

$$[X_{21},\tilde{R}_{11}]^T,\ [X_{11},\tilde{R}_1]^T,\ [X_1,\tilde{R}_{22}]^T, \text{ and } [X_2,\tilde{R}_2]^T;$$

it successively transforms them into the following nX(n/2) block-submatrices,

$$[\tilde{R}_1,0]^T,\ [R_{11},0]^T,\ [\tilde{R}_2,0]^T, \text{ and } [R_{22},0]^T$$

by premultiplication by the nXn unitary matrices U_1,U_2,U_3, and U_4, respectively. (Here we write U_k to distinguish those nXn unitary matrices from the 2X2 submatrices Q_k of the factors $\tilde{Q}_k$ in (20.2). Actually the matrices U_k and Q_k play similar roles and even coincide (within their scaling) where n is reduced to 2. Here and hereafter W^T designates the <u>transpose</u> <u>of</u> <u>a</u> <u>matrix</u> W.) The first two (of the four matrix transformations above) linearly transform the first column of the blocks of the given matrix $[X,\tilde{R}]^T$ and define the respective transformations in the second column,

$$\begin{bmatrix} X_0 \\ X_1 \end{bmatrix} = U_1 \begin{bmatrix} X_{22} \\ \tilde{R}_{22} \end{bmatrix}, \quad \begin{bmatrix} X_3 \\ X_2 \end{bmatrix} = U_2 \begin{bmatrix} X_{12} \\ X_0 \end{bmatrix} . \qquad (20.10)$$

The two subsequent linear transformations are applied to the second column of the blocks and do not require to change the corresponding 0-blocks in the first column.

The output matrices are obtained in the form,

$$R = \begin{bmatrix} R_{11} & R_{12} \\ 0 & R_{22} \end{bmatrix}, \quad \tilde{Q} = \begin{bmatrix} I_n & 0 & 0 \\ 0 & U_4 & 0 \\ 0 & 0 & I_n \end{bmatrix} \begin{bmatrix} U_2 & 0 \\ 0 & U_3 \end{bmatrix} \begin{bmatrix} I_n & 0 & 0 \\ 0 & U_1 & 0 \\ 0 & 0 & I_n \end{bmatrix}. \qquad (20.11)$$

It is easy to verify that such matrices satisfy the requirements (20.8).

Summarizing we have the following implication (mapping) of the computational problems where n is even.

$$\tilde{Q}R(n) \leftarrow 4 \circledcirc \tilde{Q}R(n/2) \circledcirc 2 \circledcirc MM(n,n,n/2) \circledcirc 12\ MM(n/2) \circledcirc MM(n),$$

that is, $\tilde{Q}R(n)$ can be solved if 4 problems $\tilde{Q}R(n/2)$, 2 problems MM(n,n,n/2), 12 problems MM(n/2), and one problem MM(n) are solved.

We may reduce each of the rectangular MM's and also MM(n) to several MM(n/2)'s and MA's, (compare Notation 20.1) and derive the following mapping (implication),

$$\tilde{Q}R(n) \leftarrow 4 \circledcirc \tilde{Q}R(n/2) \circledcirc 28 \circledcirc MM(n/2) \circledcirc 8 \circledcirc MA(n/2,n/2). \qquad (20.12)$$

Recursively applying (20.12) for $n=2^h$, $h=k,k-1,\ldots,1$ and then applying Algorithm 20.2 for n=1, we solve the problem $\tilde{Q}R(n)$ involving a total of at most n^2 evaluations of square roots (at the steps (20.9)) and of O(MM(n)) arithmetical operations. Since (20.7) implies (20.1) for $\tilde{R}=0$, this also proves Theorem 20.3. Since (20.2) is assured even if the steps (20.9) are simplified by the substitutions of R=1, this also proves Theorem 20.4. ∎

21. Applications of the QR- and the QR-Type Factorization to the Problems MI, SLE and Det.

In this section we will apply the unscaled QR-factorization (20.1) and the unscaled $\tilde{Q}R$-factorization (20.2) as two alternatives to the argument of Section 19, that is, as a means of the reduction of MI, SLE, and Det to MM. That reduction will nowhere involve divisions by zero because both QR- and QR-type factorization do not involve them. At first we will use (20.1). We will start with the case of SLE. (20.1) implies that a system

$$X\underline{z} = \underline{v} \tag{21.1}$$

of n linear equations in n unknowns for all nXn matrices X and all n-dimensional vectors $\underline{v}$ is equivalent to the triangular system

$$\Sigma R \underline{z} = Q^H \underline{v}. \tag{21.2}$$

Therefore the solution $\underline{z}$ can be evaluated by back substitution, see [A], p. 436, involving $O(n^2)$ arithmetical operations provided that the matrices Q, R, and Σ satisfying (20.1) are known, see [A], p. 436. It follows, see Theorem 20.1, that

$$ar(SLE(n)) = O(MM(n)) \text{ as } n \to \infty . \tag{21.3}$$

((21.3) follows also from (19.4) and (19.9) but here we have a different proof.) ∎

Similarly (20.1) implies that

$$Q^{-1} = \Sigma^{-1} Q^H, \; X^{-1} = R^{-1} \Sigma^{-1} Q^H = (\Sigma R)^{-1} Q^H$$

if X is a square nXn invertible matrix. This reduces the problem MI(n) as follows.

$$MI(n) \leftarrow QR^*(n) \circledast 2 \odot MM(n) \circledast TMI(n). \tag{21.4}$$

Here and hereafter TMI(n) <u>designates the problem of the inversion of an nXn triangular matrix</u>. (Actually one of the two problems MM(n) in the latter implication is the problem of multiplication of the diagonal matrix Σ by the triangular matrix R, which amounts to $(n^2+n)/2$ multiplications of the entries of Σ and R.)

Next note that, in the case where X=R is a (2n)X(2n) upper triangular matrix, we may represent that matrix as the 2X2 block matrix,

$$R = \begin{bmatrix} R_{11} & R_{12} \\ 0 & R_{22} \end{bmatrix} . \tag{21.5}$$

Here R_{11} and R_{22} are nXn upper triangular matrices. Then the matrix equation (19.6) takes the following form,

$$R^{-1} = \begin{bmatrix} R_{11}^{-1} & -R_{11}^{-1} R_{12} R_{22}^{-1} \\ 0 & R_{22}^{-1} \end{bmatrix} . \tag{21.6}$$

Note that the matrices R_{11} and R_{22} are invertible if R is invertible. (21.5) and (21.6) imply the following reduction of the computational problems.

$$\mathrm{TMI}(2n) \leftarrow 2 \odot \mathrm{TMI}(n) \circledast 2 \odot \mathrm{MM}(n) \circledast \mathrm{MS}(n,n). \tag{21.7}$$

Applying (21.7) recursively for $n = 2^h$, $h=k,k-1,\dots,0$, we derive that

$$\mathrm{ar}(\mathrm{TMI}(n)) = O(\mathrm{MM}(n)) \text{ as } n \to \infty . \tag{21.8}$$

Combining Theorem 20.1 with the relations (21.4) and (21.8), we again arrive at (19.9) using a different proof. (If the given matrix R has the size NXN where log N is not integer, then invert the $2^k X 2^k$ matrix $\tilde{R} = \begin{bmatrix} R & 0 \\ 0 & I \end{bmatrix}$ where $k = \lceil \log N \rceil$ and read R^{-1} in the upper left corner of the matrix $\tilde{R}^{-1}$.) ■

Next we recall that

$$\det (V^H) = (\det V)^H, \det(UV) = \det U \det V$$

for all square matrices U and V. The latter properties and (20.1) immediately imply that

$$\det X = \det Q \det R, \det Q (\det Q)^H = \det \Sigma$$

and therefore

$$\det X (\det X)^H = \det R (\det R)^H \det \Sigma.$$

(Recall that z^H designates the complex conjugate of the complex number z as we denoted this in Section 19.) Therefore the evaluation of det X $(\det X)^H$ for all nXn matrices X is reduced to the problem QR(n) and to at most n multiplications. ■

Next we will apply the factorization (20.2) to SLE and Det. (The application of (20.2) to MI is not efficient.) Since Q^H is a nonsingular matrix, (20.2) implies that the system (21.1) is equivalent to the system $Q^H X \underline{z} = Q^H \underline{v}$. Substitute here (20.2) and obtain the next equivalent system of linear equations,

$$R \underline{z} = \underline{u} \text{ where } \underline{u} = \tilde{Q}_{4^h} \tilde{Q}_{4^h-1} \cdots \tilde{Q}_1 \underline{v},\ n \leq 2^h < 2n. \tag{21.9}$$

Successively multiply the vector $\underline{v}$ by the matrices $\tilde{Q}_1, \tilde{Q}_2, \ldots, \tilde{Q}_{4^h}$ involving at most four multiplications and two additions for each multiplication of a vector by the matrix Q_s, $s=1,2,\ldots,4^h$. Then the vector $\underline{u}$ will be computed in at most $6*4^h < 24n^2$ arithmetical operations. Then solve the system $R \underline{z} = \underline{u}$ by back substitution using $O(n^2)$ arithmetical operations. Summarizing and taking into account Theorem 20.4 again arrive at (21.3). ■

Finally we will apply the factorization (20.2) to Det(n). As follows from (20.2),

$$\det Q^H \det X = \det R,\ \det Q^H = \prod_{s=1}^{4^h} \det \tilde{Q}_s \tag{21.10}$$

where $4^h < 4n^2$. $\det \tilde{Q}_s = \det Q_s$ equals the determinant of a 2X2 matrix and can be computed in two multiplications and one subtraction for every s. Therefore the evaluation of $\det Q^H$ costs at most $4^{h+1}-1 < 16n^2$ arithmetical operations. (Furthermore $\det Q_s = 1$ for all s if the computation by Algorithm 20.2 has been performed with scaling that makes the matrices Q_s unitary. Then also $\det Q^H = 1$ and need not be evaluated.) Only n-1 multiplications are needed for the evaluation of the determinant of the triangular matrix R.

Summarizing those estimates and taking into account Theorem 20.4, we obtain a yet another proof of (19.11). ■

22. Storage Space for Asymptotically Fast Matrix Operations.

Does the asymptotic acceleration of MM and of other matrix operations involve any substantial increase of the storage space? The negative answer for MM is given by the next simple theorem first appeared as Remark 2.3 in [P82b].

Theorem 22.1. For an arbitrary ring F and arbitrary positive ε, the N X N MM can be performed over F involving simultaneously $O(N^{\omega+\varepsilon})$ arithmetical operations and $O(N^2)$ units of storage space as $N \to \infty$, provided that each operand as well as each output of an arithmetical operation occupies one unit of storage space and that $\omega = \omega_F$ is the exponent of MM over F.

Remark 22.1. The space bound of Theorem 22.1 is sharp up to within a constant factor because $2N^2$ units of storage space are necessary already in order to store the input information.

Proof of Theorem 22.1 relies on the following observations.

(i) For some n there exists a bilinear algorithm $BA(n) = BA_F(n)$ of rank $M = n^{\omega+\varepsilon}$ for $n\text{X}n$ MM over the ring F, see Theorem 2.4. Such an algorithm can be used as a basis of recursive construction of bilinear algorithms $BA(N) = BA_F(N)$ for $N \text{ X } N$ MM that involve $O(N^{\omega+\varepsilon})$ arithmetical operations as $N \to \infty$, see the proof of Theorem 2.1 in the case m=n=p in Section 2.

(ii) Whenever a value L_qL_q' for some q is computed by the basis bilinear algorithm BA(n), such a value can be immediately included into the sums (2.3) and neither L_q, nor L_q', nor L_qL_q' need not be stored afterwards. The same observation holds for the algorithms $BA(n^h)$ for all h, where L_q, L_q', and L_qL_q' are $n^{h-1} \text{ X } n^{h-1}$ matrices.

(iii) The same $3Mn^2$ constants $f(i,j,q)$, $f'(j,k,q)$, $f''(k,i,q)$, $i,j,k=0,1,\ldots,n-1$, $q=0,1,\ldots,M-1$, are used at all steps of all recursive algorithms BA(N).

Let us supply some comments to part (ii) and some estimates for the space used by BA(N). We may assume that $N = n^h$, compare the proof of Theorem 2.1.

Represent the two given $n^{h+1} \text{ X } n^{h+1}$ matrices X and Y as two $n \text{ X } n$ matrices $X = [x_{ij}]$, $Y = [y_{jk}]$ where i,j,k range from 0 to n-1 and where the entries x_{ij}, y_{jk} are $n^h \text{ X } n^h$ block-submatrices of X and Y. Let P(k,i,0) denote the $n^h \text{ X } n^h$ zero-matrix for all k,i. Evaluate the product XY in M steps as follows.

Step q, $q=0,1,\ldots,M-1$. Successively evaluate the following $n^h \text{ X } n^h$ matrices, compare (2.2),(2.3). $L_q = \sum_{i,j} f(i,j,q)x_{ij}$, $L_q' = \sum_{j,k} f'(j,k,q)y_{jk}$, L_qL_q', $P(k,i,q+1) = P(k,i,q) + f''(k,i,q)\, L_qL_q'$. (Then the outputs of Step M-1 are just the desired values $\sum_j x_{ij}y_{jk} = P(k,i,M)$ for all k,i.) Require that the evaluation of the matrix products L_qL_q' be performed by the algorithm $BA(n^h)$ and use the straightforward way for other operations of Step q. ∎

It remains to allocate and to estimate the storage space in the above algorithm $BA(n^{h+1})$. We will do that as follows.

Store the $2n^{2h+2}$ entries of x_{ij}, y_{jk} and the $3Mn^2$ constants $f(i,j,q)$, $f'(j,k,q)$, $f''(k,i,q)$ for all i,j,k,q throughout the computation. In addition, allow $W(n^{h+1})$ additional units of space that constitute the working array for $n^{h+1} \text{ X } n^{h+1}$ MM performed by the algorithm $BA(n^{h+1})$ where $W(n^{h+1})$ is defined by the following recursive equations,

$$W(n^{h+1}) = (n^2 + 2)\, n^{2h} + W(n^h),\ h=0,1,\ldots,;\ W(1)=0. \tag{22.1}$$

Allocate the space for the working array as follows.

(i) Use n^{2h+2} units for the storage of the $n^h \times n^h$ matrices $P(k,i,q)$ for all i,k,q. (When the matrix $P(k,i,q+1)$ has been evaluated, it replaces the matrix $P(k,i,q)$. When Step q has been completed, there is no need to store the matrices $L_q L_q'$, $P(k,i,q)$ anymore.)

(ii) Use $2n^{2h}$ units in order to evaluate and to store the $n^h \times n^h$ matrices L_q, L_q'. (There is no need to store the matrices L_q and L_q' anymore when their product $L_q L_q'$ has been evaluated.)

(iii) Use $W(n^h)$ units as the working array for the evaluation of $L_q L_q'$ for all q.

To estimate $W(n^{h+1})$, successively substitute $h-1, h-2, \ldots, 0$ for h in (22.1). Combine the h+1 resulting inequalities. Finally substitute $W(1)=0$ and obtain that

$$W(n^{h+1}) \leq (n^2 + 2) \sum_{g=0}^{h} n^{2g} = (n^2 + 2) \frac{n^{2h+2}-1}{n^2-1} < 2n^{2h+2}$$

since $n>1$. We surely may assume that $M \leq n^3$. Then a total of less than $4n^{2h+2} + 3n^5 = (4 + 3n^{3-2h})n^{2h+2}$ units of storage space suffice in our algorithm $BA(n^{h+1})$ for $n^{h+1} \times n^{h+1}$ MM for arbitrary h. ■

Remark 22.2. Theorem 22.1 is immediately extended to the sharp (up to within a constant factor) upper bounds $O(N^2)$ as $N \to \infty$ on the number of units of storage space required in the asymptotically fast algorithms of the previous sections for BMM(N), MI(N), Det(N), and SLE(N). (See [Ke] about the problem CCP.) We leave the verification of those simple extensions as an exercise for a reader.

23. The Bit-Complexity of Computations in Linear Algebra. The Case of Matrix Multiplication.

In the remaining sections of this part of the monograph we will derive some upper bounds on the numbers of bit-operations required in the algorithms for MM(n), QR(n), SLE(n), MI(n), and Det(n). Our study will also lead to estimating the numbers of bits of storage space required in those algorithms and will give a quantitative characterization of their stability.

We will work with algorithms that involve only arithmetical operations (and the evaluation of square roots in the case of the algorithms for QR(n)). We already have some bounds on the number of such operations (arithmetics and square roots) required in order to solve the listed computational problems. It remains to combine those bounds with the estimates for the number of binary bits that we need to use in the operands. When the latter estimates are obtained, it will remain to apply Theorems 18.2-18.5 and their simple extensions to the case where the arithmetical operations and the evaluation of square roots are performed over s-bit binary numbers rather than over integers. The restriction on the number of bits in the representation of the operands and outputs of the arithmetical operations implies that the computation is performed approximately. We will impose certain bounds on the magnitudes of the absolute errors in the final outputs and will minimize the number of bits that must be involved under that condition. Such an approach is customary in the study of the evaluation of (sets of) polynomials, rational functions, and other algebraic expressions, see [Tr]. In this section we will consider a simpler case where we will deal with MM, compare [Br70],[R80],[P81a]. We will estimate the numbers of bit-operations and of bits of storage space that suffice in order to evaluate the approximations z^*_{ik} to $z_{ik} = \sum_j x_{ij}y_{jk}$ in the domain $D = D(\mu'(n), \mu''(n))$ such that

$$|x_{ij}| \leq \mu'(n); |y_{jk}| \leq \mu''(n) \text{ for all } i,j,k \tag{23.1}$$

with the precision $e(n)$ such that

$$|z^*_{ik} - \sum_j x_{ij}\, y_{jk}| \leq e(n) \text{ for all } i,k. \tag{23.2}$$

Here i,j,k range from 0 to $n-1$; $\mu'(n),\mu''(n),e(n)$ are positive parameters that depend only on n and that satisfy the inequality

$$n\,\mu'(n)\mu''(n) > e(n). \tag{23.3}$$

(If (23.3) does not hold, then (23.2) is satisfied in the domain D for $z^*_{ik} = 0$ for all i,k.) For simplicity, let the variables and the constants be real and compare Remark 23.3 below. We will apply $BA(n)$, an arbitrary bilinear algorithm (2.2),(2.3) for nXn MM (which may involve arbitrary real constants), represent all inputs and all computed values as normalized floating point binary numbers, and perform all arithmetical operations of the algorithm with chopping the fractions (mantissas) of all operands and outputs to a certain number $s=s(D,e(n))$ of binary bits. Hereafter such chopping will be called <u>chopping to</u> s <u>(binary) bits throughout the computation</u> (or <u>throughout the algorithm</u>). The choice of s must be such that (23.2) holds in the domain D.

Hereafter we will rely on the following customary assumption.

<u>Assumption 23.1</u>. The operations over the exponents of the floating point binary numbers have negligible cost (because the cost of the operations over the fractions of those numbers strongly dominates); furthermore Theorem 18.2-18.5 can be extended from the case of the operations over integers modulo 2^s to the case of the operations over binary numbers having s bits in their normalized floating point fractions, compare [K].

We will apply the following simple observation.

<u>Proposition 23.1</u>. Under Assumption 23.1, the following upper estimate holds for the <u>bit-time</u> $bt(A)$, that is, for the number of bit-operations involved in an algorithm A for MM(n), SLE(n), MI(n), or Det(n) (applied with chopping to s bits throughout the computation), $bt(A) \leq ar(A)\, bt(s)$. Here $ar(A)$ is the number of

arithmetical operations in A and bt(s) is defined by (18.2),(18.3). Simultaneously $bs(A) = O(n^2 s)$ as $n \to \infty$ where bs(A) is the bit-space of A , that is, the number of bits of storage space, compare Theorem 22.1 and Remark 22.2.

The desired upper estimates (for the bit-time and the bit-space required for the approximation to XY in the domain D with the precision e(n) by a given algorithm A with chopping to s bits throughout the computation) have now been reduced to estimating an upper bound on s such that (23.2) holds in the domain D. Next we will solve that problem by estimating the errors in D as functions in s. We will use the following notation.

Notation 23.1. $h(s) = 2^{1-s}$ denotes the unit chopping error, which is the maximum relative error caused by chopping the floating point fraction of a real nonzero binary number u to s bits, that is, if u(s) designates the chopped value, then $\max_u |u(s)-u)/u| = h(s)$, compare [CdB], pp. 8-9. We will also write $e(A,D,h(s)) = \text{maximum } |z^*_{ik} - \sum_j x_{ij}y_{jk}|$ where z^*_{ik} are the approximations to $\sum_j x_{ij}y_{jk}$ computed by the algorithm A with chopping to s bits throughout the computation and where the maximum is in all i,k and in all inputs X and Y in D.

The following simple result can be found in [CdB], pp. 8-11.

Proposition 23.2. Let b and t be the approximations to $\sum_{k=0}^{K-1} u_k v_k$ and $\sum_{k=0}^{K-1} u_k v_k w_k$, respectively, computed by the straightforward algorithms with chopping the floating point fractions of all inputs and of all computed values to s binary bits (that is, with chopping to s bits throughout the computation). Let $h(s) = 2^{1-s}$, see Notation 23.1; $s \to \infty$, $h(s) \to 0$. Then

$$|b - \sum_k u_k v_k| \le \sum_k ((1+h(s))^{K-k} - 1)\, |u_k v_k| \le h(s)K \sum_k |u_k v_k| + O(h^2(s)),$$

$$|t - \sum_k u_k v_k w_k| \le \sum_k ((1+h(s))^{K+1-k} - 1)|u_k v_k w_k| = h(s)(K+1) \sum_k |u_k v_k w_k| + O(h^2(s)).$$

Next assume that the matrix product XY has been computed by the given bilinear algorithm (2.2),(2.3) with chopping to s bits throughout the computation and let us estimate the resulting error using Proposition 23.1.

Hereafter we will ignore the values of the order $O(h^2(s))$. We will designate

$$f = \max_{i,j,q} |f(i,j,q)|,\quad f' = \max_{j,k,q} |f'(j,k,q)|,\quad f'' = \max_{k,i,q} |f''(k,i,q)|. \qquad (23.4)$$

Theorem 23.3. Let $\mu'(n)$ and $\mu''(n)$ be two positive parameters depending only on n. Let a bilinear algorithm BA(n) (see (2.2),(2.3)) of rank M=M(n) be applied to the evaluation of an nXn matrix product $XY = [\sum_j x_{ij}y_{jk}]$ with chopping the floating point fractions to s binary bits throughout the computation. Then the computed outputs z^*_{ki} approximate to $\sum_j x_{ij}y_{jk}$ in the domain D (defined by (23.1)) with the precision (with the bound on the magnitudes of the absolute output errors)

$$e(n) = e(A,D,h(s)) \leq h(s)M(M+2n^2+1)n^4 f\, f' f'' \mu'(n)\mu''(n) + O(h^2(s)),$$

compare Notation 23.1. Here f, f', and f'' are defined by (23.4).

Proof. Let L_q^* and $L_q'^*$ designate the computed approximations to L_q and L_q', respectively. At first we will apply Proposition 23.2 in order to estimate the errors $\Delta L_q = L_q^* - L_q$ of the evaluation of $L_q = \sum_{i,j} f(i,j,q)x_{ij}$ (where in+j, n^2, $f(i,j,q)$, x_{ij} play the roles of k, K, u_k, v_k, respectively). Ignoring the values $O(h^2(s))$ we will derive that

$$|\Delta L_q| \leq h(s)n^2 \sum_{i,j} |f(i,j,q)|*|x_{ij}| \leq h(s)n^4 f\, \mu'(n),$$

compare (23.1),(23.4).

Similarly we will estimate that for all q

$$|L_q'^* - L_q'| = |\Delta L_q'| \leq h(s)n^4 f' \mu''(n).$$

Therefore for all q,

$$|L_q^* L_q'^* - L_q L_q'| \leq |L_q^* \Delta L_q' + (\Delta L_q) L_q'| \leq$$

$$h(s)n^4(|L_q^*| f' \mu''(n) + |L_q'| f\, \mu'(n)) \leq h(s)n^4(|L_q| f' \mu''(n) +$$

$$|L_q'| f\, \mu'(n)||) + h(s)n^4 f' \mu''(n) |\Delta L_q| \; .$$

Ignoring the values

$$h(s)n^4 f' \mu''(n) |\Delta L_q| \leq h^2(s)n^8 f f' \mu'(n)\mu''(n)$$

of the order $h^2(s)$ and substituting the obvious bounds

$$|L_q| \leq n^2 f\, \mu'(n), \quad |L_q'| \leq n^2 f' \mu''(n), \tag{23.5}$$

compare (23.1),(23.4), we derive that for all q

$$|L_q^* L_q'^* - L_q L_q'| \leq 2h(s)n^6 f\, f' \mu'(n)\mu''(n).$$

Consequently for all k,i

$$|\sum_q f''(k,i,q) L_q^* L_q'^* - \sum_j x_{ij} y_{jk}| \leq \tag{23.6}$$

$$\sum_q |f''(k,i,q)|*|L_q^* L_q'^* - L_q L_q'| \le 2h(s)Mn^6 f\, f' f'' \mu'(n)\mu''(n),$$

It remains to estimate the error Δ_{ki}^* of the evaluation of $\sum_q f''(k,i,q)L_q^* L_q'^*$ with chopping to s bits throughout. We may apply Proposition 23.2 again (where $q,M,f''(k,i,q),L_q^*,L_q'^*$ play the roles of k,K,u_k,v_k,w_k, respectively). Ignoring the values of the order $h^2(s)$ we deduce that

$$|\Delta_{ki}^*| \le h(s)(M+1) \sum_q |f''(k,i,q)L_q^* L_q'^*|$$

and also extend (23.5) to the upper bounds on $|L_q^*|$ and $|L_q'^*|$. Substitute those bounds, apply (23.4), and obtain that for all k,i

$$|\Delta_{ki}^*| \le h(s)(M+1)M\, n^4 f\, f' f'' \mu'(n)\mu''(n).$$

Combine the latter inequality with (23.6) and derive Theorem 23.3. ∎

In the next corollary we will estimate the asymptotic error bounds as $n \to \infty$.

Corollary 23.4. Under the assumptions of Theorem 23.3, let BA(n) be a recursive bilinear algorithm for nXn MM (such as ones defined in the proofs of Theorems 2.1 and 22.1). Let $M(n) \le n^3$. Then the output error bound satisfies the following estimate,

$$e(n) = e(A,D,h(s)) = h(s)\ O(n^{13})\mu'(n)\mu''(n) + O(h^2(s))$$

as $n \to \infty$ and $s=s(n) \to \infty$. Here $h(s) = 2^{1-s}$, compare Notation 23.1.

Proof. We only need to note that $f=O(n)$, $f' = O(n)$, $f'' = O(n)$ as $n \to \infty$, which immediately follows from (2.12). ∎

Next we will choose s such that (23.2) holds in the domain (23.1), compare Assumption 23.1.

Corollary 23.5. Under the assumptions of Corollary 23.4, it is sufficient to choose

$$s = \lceil \log(2M(M+2n^2+1)n^4 f\, f' f'' \mu'(n)\mu''(n)/e(n)) \rceil + O(2^{-s}) = \qquad (23.7)$$

$$O(\log(n^{13}\mu'(n)\mu''(n)/e(n))) + O(1)$$

($s \to \infty$ as $n \to \infty$) in order to satisfy the error bound (23.2) in the domain (23.1) (such that (23.3) holds) by applying the algorithm BA(n) with chopping to s bits throughout the computation. Such a computation simultaneously involves only bt(BA(n))=ar(BA(n))bt(s) bit-operations and bs(BA(n))=$O(n^2 s)$ bits of storage space where s is defined above, bt(s) is defined by (18.2),(18.3), and ar(BA(n)) is the number of arithmetical operations involved in BA(n), compare Proposition 23.1.

Choosing asymptotically fast bilinear algorithm BA(n) (compare Theorem 2.4) we obtain the following result.

Theorem 23.6. The product XY of a pair of nXn matrices X and Y can be evaluated in the domain (23.1) with the error bound (23.2) (such that (23.3) holds) involving simultaneously

$$bt(MM(n))=O(n^{\omega+\varepsilon}bt(s)).$$

bit-operations and $bs(MM(n)) = O(n^2 s)$ bits of storage space. Here s is defined by (23.7), ω is the exponent of MM over the field of real constants, and ε is arbitrary positive number.

Remark 23.1. Theorem 23.6 enables us to extend Theorem 18.6 to Theorem 18.7. (Choose $e(n) < 0.5$.)

Remark 23.2. Another approach to estimating the bit-complexity of bilinear algorithms of MM (over the ring of integer constants) relies on the application of modular arithmetic, see [P82]. In that case it suffices to choose $s = 6+2\log(n^{1.5}\mu'(n)\mu''(n)/e(n)))$ in order to assure (23.2) in D. This means roughly double increase of s against (23.7) if $\log(n^{1.5}\mu'(n)\mu''(n)/e(n))$ dominates over $\log(2M(M+2n^2+1)n^{2.5}f f'f'')$. Such an approach can be extended to estimating the bit-operation complexity of other bilinear problems, in particular, of polynomial multiplication, see [G] and [P82] and compare [P81c]. It can be also extended to bilinear algorithms that involve rational constants.

Remark 23.3. Our estimates for the bit-complexity are immediately extended from the case where the constants of the given algorithms and the given matrices X and Y are real to the case where they are complex because each arithmetical operation over complex numbers can be immediately reduced to few arithmetical operations over real numbers. For a similar reason, we will estimate the bit-complexity of other problems also only in the real case.

Finally we will extend our estimates to the case of bilinear λ-algorithms of positive degree d. It is obvious that the asymptotic bounds on the bit-time and on the bit-space remain the same as in the case of conventional bilinear algorithms, as follows from Theorem 6.3 and from its proof, but, rather surprisingly, even the estimates for the bit-time of nXn MM for specific n increase only about d+1 times if we apply the original idea of [BCLR] about the approximation by choosing small λ.

Theorem 23.7. Let $BA(n,\lambda)$ be a given bilinear λ-algorithm (6.3),(6.4) of degree d and of λ-rank M for nXn MM and let its constants be represented as polynomials of degree at most d in λ,

$$f(i,j,q,\lambda) = \sum_{g=0}^{d} \tilde{f}(i,j,q,g)\lambda^g, f'(j,k,q,\lambda) = \sum_{g=0}^{d} \tilde{f}'(j,k,q,g)\lambda^g, \quad (23.8)$$

$$f''(k,i,q,\lambda) = \sum_{g=0}^{d} \tilde{f}''(k,i,q,g)\lambda^g.$$

Let $E(n)$, $\mu'(n)$, $\mu''(n)$ be positive parameters that satisfy (23.3). Let

$$\left.\begin{array}{l} \tilde{f} = \underset{i,j,q,g}{\text{maximum}} |\tilde{f}(i,j,q,g)|, \ \tilde{f}' = \underset{j,k,q,g}{\text{maximum}} |\tilde{f}''(j,k,q,g)|, \\ \tilde{f}'' = \underset{k,i,q,g}{\text{maximum}} |\tilde{f}''(k,i,q,g)|, \end{array}\right\} \quad (23.9)$$

$$-\log \lambda = \lceil \log(2(d+1)^3 M n^4 \tilde{f}\ \tilde{f}'\tilde{f}''\ \mu'(n)\mu''(n)/e(n))\rceil, \quad (23.10)$$

$$s = \lceil \log(4(d+1)^3 M(M+2n^2+1)n^4\tilde{f}\ \tilde{f}'\tilde{f}''\mu'(n)\mu''(n)/e(n))\rceil - d \log \lambda + O(2^{-s}). \quad (23.11)$$

Then the errors of the approximations computed by the given λ-algorithm (for λ defined by (23.10) and with chopping to s binary bits throughout the computation) will satisfy (23.2) in the domain (23.1). The computation involves only $bt(BA(n,\lambda)) = ar(BA(n,\lambda))bt(s)$ bit-operations and simultaneously only $bs(BA(n,\lambda)) = O(n^2 s)$ bits of storage space, compare Proposition 23.1. Here $ar(BA(n,\lambda))$ denotes the number of all arithmetical operations involved in $BA(n,\lambda)$; $bt(s)$ is defined by (18.2),(18.3).

Remark 23.4. The upper bound on $bt(BA(n,\lambda))$ can be decreased because every multiplication (division) by λ amounts to a simpler operation of shifting the radix point of a binary number since λ is a power of 2, see (23.10).

Proof. At first we will assume that $s = \infty$, that is, that all arithmetical operations have been performed with the infinite precision and that the output errors stem only from using λ-algorithm (6.3),(6.4). Let us estimate that error.

Substitute (23.8) into (6.3),(6.4) and deduce that

$$|\Delta_{ik}| = |\sum_q f''(k,i,q,\lambda)L_q L'_q - \sum_j x_{ij}y_{jk}| \leq (d+1)^3 \lambda\ n^4 M\ \tilde{f}\ \tilde{f}'\tilde{f}''\mu'(n)\mu''(n)$$

for all i,k. Here $\tilde{f},\tilde{f}',\tilde{f}''$ are defined by (23.9) and we assume that $0 < \lambda \leq 1$.

Therefore it suffices to define λ by (23.10) in order to assure that

$$|\Delta_{ik}| \leq e(n)/2 \text{ for all } i \text{ and } k.$$

It remains to choose s such that the error caused by chopping to s bits throughout the computation will not exceed $e(n)/2$. This can be done as in the proof of Theorem 23.3 applied to the bilinear algorithm obtained from the given λ-algorithm where λ is defined by (23.10). In that bilinear algorithm the constants are defined by the constants of the λ-algorithm BA(n,λ) as follows,

$$f(i,j,q) = f(i,j,q,\lambda),\ f'(j,k,q) = f'(j,k,q,\lambda),\ f''(k,i,q) = \lambda^{-d} f''(k,i,q,\lambda)$$

for all i,j,k,q and for λ defined by (23.10). We will assume that $0 < \lambda \leq 1$. Then

$$|f(i,j,q)| \leq (d+1)\tilde{f}, |f'(j,k,q)| \leq (d+1)\tilde{f}', |f''(k,i,q)| \leq (d+1)\ \lambda^{-d}\tilde{f}'' \text{ for all } i,j,k,q$$

and therefore $f\,f'f'' \leq \lambda^{-d}\tilde{f}\,\tilde{f}'\tilde{f}''(d+1)^3$. Then repeating the error analysis of the proof of Theorem 23.3, we will verify that the choice of s in (23.11) does guarantee the upper bound e(n)/2 on the errors caused by chopping. ∎

Finally we note that in this section we estimated the bit-time and the bit-space sufficient in order to evaluate an approximation in a given domain D with a prescribed precision E to the nXn matrix product XY by an algorithm A=BA(n) and by a λ-algorithm A=BA(n,λ) applied with chopping to s binary bits throughout the computation (for a certain choice of s). Hereafter bt(A,D,E) and bs(A,D,E) will designate such <u>bit-time</u> and <u>bit-space</u>, respectively. Similar notation will be applied to the algorithms A for other arithmetical computational problems, such as SLE, MI, Det. If Z(V) is the output array (matrix) of such problem we will denote

$$bt(Z(V),D,E) = \underset{A}{\text{minimum}}\ bt(A,D,E), bs(Z(V),D,E) = \underset{A}{\text{minimum}}\ bs(A,D,E) \qquad (23.12)$$

where A ranges among all arithmetical algorithms for the evaluation of Z(V).

24. Matrix Norms and Their Application to Estimating the Bit-Complexity of Matrix Multiplication.

Estimating the bit-complexity of the problems MI, SLE, and Det, we will use the norms of matrices and vectors. In this section we will recall the customary definitions and some simple properties of the norms, see [A], pp. 412-420; [FF], § 13; [GvL], pp. 14-16, and will express the estimates of the previous section in terms of the norms of the matrices formed by the inputs and by the errors.

<u>Definition 24.1</u>. A nonnegative function $||W||$ defined on the set of all square and rectangular matrices {W} is called <u>a matrix norm</u> if the following conditions are met.

$$||W|| = 0 \text{ if and only if } W = 0; \qquad (24.1)$$

$$||cW|| = |c| * ||W|| \text{ for all } W \text{ and for all complex constants } c; \qquad (24.2)$$

$$||W + V|| \leq ||W|| + ||V|| \text{ and } ||WV|| \leq ||W|| * ||V|| \qquad (24.3)$$

where the sizes of matrices W and V are such that their additions and/or multiplications are defined. If the requirement that $||WV|| \leq ||W|| * ||V||$ has been waived and if the norm has been defined only on the set of all column-vectors, that is, on the set of m X 1 matrices, then such a norm is called a vector norm. A matrix norm and a vector norm are called compatible if

$$||W\underline{v}|| \leq ||W|| * ||\underline{v}|| \tag{24.4}$$

for all matrices W and vectors $\underline{v}$ whose products $W\underline{v}$ are defined.

Example 24.1, the ℓ_h- norms of vectors and matrices, $0 < h \leq \infty$. The ℓ_h-norm of a vector $\underline{v} = [v_j]$ equals

$$||\underline{v}||_h = (\sum_j |v_j|^h)^{1/h} \text{ if } 0<h<\infty, \; ||\underline{v}||_\infty = \underset{j}{\text{maximum}}|v_j|.$$

In particular $||\underline{v}||_2 = (\underline{v}^H\underline{v})^{1/2}$. The ℓ_h-norm of a matrix W equals

$$||W||_h = \underset{\underline{v}\neq 0}{\text{maximum}}(||W\underline{v}||_h \;/\; ||\underline{v}||_h).$$

The relations (24.1)-(24.4) can be easily verified for the ℓ_h-norms for all h, compare [A], pp. 412-420. In the sequel we will rely on the ℓ_h-norm for h=2 and h=∞. We will simplify our notation by writing $||W||$ at the place of $||W||_h$ for h=2 and h= ∞, that is, in the cases where any of the two norms (ℓ_2 or ℓ_∞) can be applied. Those two norms have the following simple and well-known properties, compare [A], pp. 416-420; [FF], § 13; [GvL], pp. 14-16.

Proposition 24.1. For an arbitrary mXn matrix $W = [w_{ij}]$ and its arbitrary submatrix V (in particular V can be a 1X1 submatrix of W), the following relations hold.

$$n^{-1/2}\, ||W||_\infty \leq ||W||_2 \leq m^{1/2}\, ||W||_\infty, \tag{24.5}$$

$$\left.\begin{aligned} &||V||_\infty \leq ||W||_\infty = \underset{i}{\text{maximum}} \sum_j |w_{ij}|, \\ &||V||_2 \leq ||W||_2 = ||W^H||_2 \leq F(W) = \{\sum_{i,j} |w_{ij}|^2\}^{1/2}, \end{aligned}\right\} \tag{24.6}$$

$$||I||_\infty = ||I||_2 = 1. \tag{24.7}$$

Furthermore, for an arbitrary vector $\underline{v}$,

$$||Q\underline{v}|| = ||\underline{v}||, \; ||Q|| = ||Q^H|| = 1 \text{ if } Q^HQ = I. \tag{24.8}$$

are defined by the constants of the λ-algorithm BA(n,λ) as follows,

$$f(i,j,q) = f(i,j,q,\lambda),\ f'(j,k,q) = f'(j,k,q,\lambda),\ f''(k,i,q) = \lambda^{-d} f''(k,i,q,\lambda)$$

for all i,j,k,q and for λ defined by (23.10). We will assume that $0 < \lambda \leq 1$. Then

$$|f(i,j,q)| \leq (d+1)\tilde{f}, |f'(j,k,q)| \leq (d+1)\tilde{f}', ||f''(k,i,q)| \leq (d+1)\lambda^{-d}\tilde{f}'' \text{ for all } i,j,k,q$$

and therefore $f\,f'f'' \leq \lambda^{-d}\tilde{f}\,\tilde{f}'\tilde{f}''(d+1)^3$. Then repeating the error analysis of the proof of Theorem 23.3, we will verify that the choice of s in (23.11) does guarantee the upper bound e(n)/2 on the errors caused by chopping. ■

Finally we note that in this section we estimated the bit-time and the bit-space sufficient in order to evaluate an approximation in a given domain D with a prescribed precision E to the nXn matrix product XY by an algorithm A=BA(n) and by a λ-algorithm A=BA(n,λ) applied with chopping to s binary bits throughout the computation (for a certain choice of s). Hereafter bt(A,D,E) and bs(A,D,E), will designate such <u>bit-time</u> and <u>bit-space</u> respectively. Similar notation will be applied to the algorithms A for other arithmetical computational problems, such as SLE, MI, Det. If Z(V) is the output array (matrix) of such problem we will denote

$$\text{bt}(Z(V),D,E) = \underset{A}{\text{minimum}}\ \text{bt}(A,D,E), \text{bs}(Z(V),D,E) = \underset{A}{\text{minimum}}\ \text{bs}(A,D,E) \qquad (23.12)$$

where A ranges among all arithmetical algorithms for the evaluation of Z(V). Finally, if the matrix W^* is obtained from a matrix W by chopping the fractions in the binary representation of some of or all of the entries of W, then

$$||W^*||_\infty \leq ||W||_\infty . \qquad (24.9)$$

Next we will modify the requirements (23.1) -(23.3) using the matrix norms as follows,

$$||X|| \leq \mu'(n),\ ||Y|| \leq \mu''(n), \qquad (24.10)$$

$$||\Delta(XY)|| \leq E(n), \qquad (24.11)$$

$$\mu'(n)\,\mu''(n) > E(n) \qquad (24.12)$$

(here $\Delta(XY)$ is the error matrix$[\Delta(i,k)]$, $\Delta(i,k) = z^*_{ik} - \sum_j x_{ij}y_{jk}$, z^*_{ik} is the computed approximations to $\sum_j x_{ij}y_{jk}$, $i,k=0,1,\ldots,n-1$; $\mu'(n)$, $\mu''(n)$, and E(n) are positive parameters), and will extend the results of the previous section as follows.

Theorem 24.2. Theorems 23.3,23.6,23.7 and Corollaries 23.4,23.5 will still hold if the inequalities (24.10)-(24.12) substitute for (23.1)-(23.3) and if $E(n)/n$ substitutes for $e(n)$.

Proof. (24.6) and (24.10) imply (23.1). Therefore the cited theorems and corollaries can be extended to the domain (24.10). It remains to note that (23.2) and (24.6) imply that $\|\Delta(XY)\| \leq n\, e(n)$. ∎

To close the discussion on the bit-complexity of MM, here are the simple informational lower bounds due to the observation that we need represent the normalized fraction of each of the n^2 entries of the output matrix with at least $\log(\mu'(n)\mu''(n)/E(n))$ binary bits in order to assure the precision $E(n)$ in the domain D defined by (24.10).

Theorem 24.3. Let $E(n)$, $\mu'(n)$, $\mu''(n)$ be positive parameters; let D be defined by (24.10). Then, compare (23.12),

$$bt(XY,D,E(n)) \geq n^2\, bt(\log(\mu'(n)\mu''(n)/E(n)))$$

$$bs(XY,D,E(n)) \geq n^2 \log(\mu'(n)\mu''(n)/E(n)).$$

Remark 24.1. The lower bounds of Theorem 24.3 hold even if we replace the requirement that the fractions of all inputs, outputs, and computed values be chopped to the same number s of binary bits by a much weaker requirement that the number of bits in the fractions of each input and/or each output be invariant in the choice of inputs within the domain D and be defined only by D and $E=E(n)$.

25. Stability and Condition of Algebraic Problems and of Algorithms for Such Problems.

In this section we will define the stability and the condition of (i) algebraic computational problems such as MM, SLE, MI, and Det and of (ii) algorithms for such problems. We will see that the upper bounds on the bit-complexity and on the condition can be deduced in a single process, so that the estimates of Sections 23 and 24 immediately lead to some upper bounds on the condition of MM and of the algorithms for MM. Also we will see that the acceleration of MM does not require any noticeable sacrifice in stability (compare [Mi] where an attempt was made to seek for the opposite effect) and that even the computation by λ-algorithms of degree d for small λ requires only relatively small increase of the number of binary bits involved (that is, the increase only about d+1 times), compare the discussion in [B82]. We will start with introducing some notation. Those readers who are not interested in the study of the condition and stability may skip this section.

Definition 25.1. Hereafter a matrix V and a matrix function Z(V) designate the two arrays of the input variables and of the output values of a computational problem, respectively. W(s) denotes the matrix obtained from a matrix W by chopping the floating point fractions of the entries of W to s binary bits. In particular if W=Z(V), then Z(V,s) designates W(s), that is, the matrix obtained from the matrix Z(V) by the above chopping. (Note that generally $Z(V,s) \neq Z(V(s))$.) Let V be the input matrix and Z(V) be the output matrix of an algorithm A, that is, let A compute Z(V) from V. Let chopping to s bits of the floating point fractions of all inputs and of all computed values have been used throughout the computation by the algorithm A. Then Z(V,A,s) designates the resulting output matrix. (Note that generally $Z(V,A,s) \neq Z(V(s))$, $Z(V,A,s) \neq Z(V,s)$.) If V and ΔV are two matrices (ΔV is the perturbation matrix of the input matrix V), then $\Delta Z(V)$ will designate the matrix $Z(V+\Delta V) - Z(V)$ (of the resulting output perturbation). We will use the following notation for the perturbation caused by chopping inputs, by chopping outputs, and by chopping throughout an algorithm A, respectively,

$$\Delta(V,s) = V(s) - V,\ \Delta(Z(V),s) = Z(V,s) - Z(V),$$

$$\Delta Z(V,s) = Z(V(s)) - Z(V),\ \Delta(Z(V),A,s) = Z(V,A,s) - Z(V). \blacksquare$$

The above notation can be applied to the matrices of arbitrary sizes mXn. In particular, for n=1 and m=n=1, such matrices turn into column-vectors and into single parameters, respectively.

In the sequel we will always choose a certain matrix-vector norm (ℓ_2 or ℓ_∞) for measuring the perturbation of the input and of the output matrices. If the inputs or the outputs of the problem form two or several matrices (as in the case of the inputs to MM and of the outputs of the QR-problem), then the magnitude of the total perturbation can be measured by the sum of the norms of the two or several perturbation matrices. We will keep writing $h(s) = 2^{1-s}$, see Notation 23.1.

Next we will define the conditions of a problem and of an algorithm, will comment on their meaning, and then will modify their definitions.

Definition 25.2.

$$\text{cond}^*(Z(V),V) = \underset{\Delta V \neq 0}{\text{maximum}} \left(\frac{||\Delta Z(V)||}{||Z(V)||} \Big/ \frac{||\Delta V||}{||V||}\right),$$

$$\text{cond}(Z(V),V,h(s)) = \underset{||\Delta V|| \leq h(s)}{\text{maximum}} \frac{||\Delta Z(V)||}{||Z(V(s))||} \Big/ h(s),$$

$$\text{cond}(A,V,h(s)) = \underset{||\Delta V|| \leq h(s)}{\text{maximum}} \frac{||\Delta (Z(V),A,s)||}{||Z(V(s))||} \Big/ h(s),$$

$$\text{cond}(Z(V),V) = \lim_{s \to \infty} \text{cond}(Z(V),V,h(s)), \ \text{cond}(A,V) = \lim_{s \to \infty} \text{cond}(A,V,h(s)).$$

Comments. $\text{cond}^*(Z(V),V))$ represents the condition of the evaluation of $Z(V)$ at V, compare [A], pp. 26-28; [DB], p. 54; [Ric], pp. 46-47. $\text{cond}^*(Z(V),V)$ is equal to the maximum quotient of the norms of the relative errors of outputs and inputs caused by the perturbation of the inputs by ΔV. The maximum is in all nonzero perturbations. If the class of perturbations is restricted to the perturbations caused by chopping of inputs to s binary bits, then $\text{cond}^*(Z(V),V)$ turns into $\text{cond}(Z(V),V,h(s))$ so that $\text{cond}(Z(V),V,h(s))) \leq \text{cond}^*(Z(V),V)$. If the output errors have been caused by chopping to s bits throughout the algorithm A, then we arrive at $\text{cond}(A,V,h(s))$. In the two latter cases we use the value $h(s)$ in order to represent the relative input error (which is an upper bound on the relative error of inputs caused by chopping). The problems $Z(V)$ and algorithms A, for which $\text{cond}(Z(V),V)$ and $\text{cond}(A,V)$ are not large at V, are (rather informally) called well-conditioned at V. Otherwise they are (informally) called ill-conditioned at V.

It is assumed that the condition could serve as a qualitative measure of stability, so that the infinite value of the condition implies instability. Formally the problem of the evaluation of $Z(V)$ is called stable where $Z(V)$ is a continuous matrix function in V. An algorithm A for the evaluation of $Z(V)$ is called stable if $\lim_{s \to \infty} Z(V,A,s) = Z(V)$. The next example shows that the assumption that "extremely ill-conditioned = unstable" does not always hold in the case of MM (similar examples can be presented for MI, SLE, and Det).

Example 25.1. The above definitions of stability imply that the problem of the evaluation of matrix products XY is stable at all inputs X,Y but, on the other hand, its condition is infinite at X,Y such that, say, $||X|| \neq 0$, $||Y|| \neq 0$, $XY = 0$. (Here V is the pair X,Y and $Z(V)=XY$.) A similar discrepancy takes place if we apply Definitions 25.1 and 25.2 to the straightforward algorithm for the evaluation of XY at such X and Y.

Thus we should impose some additional requirements such as

$$||Z(V)|| \geq \eta, \ \eta \text{ is a positive constant,} \qquad (25.1)$$

in order to reconcile the concepts of ill-conditioning and instability.

To resolve the problem without imposing the bound (25.1), we may define the conditions of a problem and of an algorithm in a <u>domain of the range of variables</u> rather than at a fixed set of their values.

<u>Definition 25.3</u>. Let for two given positive values μ and ν (which could be mutually related) a matrix function $Z(V)$ be defined in the following domain $D = D(\mu, \nu)$,

$$||V|| \leq \mu, \ ||Z(V)|| \leq \nu. \qquad (25.2)$$

Let

$$\left.\begin{aligned} E(Z(\underline{V}),D,h) &= \max_{V \in D, ||\Delta V|| \leq h\mu} ||\Delta Z(V)||, \\ E(Z(V),D,h(s)) &= \max_{V \in D} ||\Delta Z(V,s)||, \\ E(A,D,h(s)) &= \max_{V \in D} ||\Delta(Z(V),A,s)||, \end{aligned}\right\} \qquad (25.3)$$

Then we will designate

$$\text{cond}(Z(V),D,h) = (E(Z(V),D,h)/\nu)/h = E(Z(V),D,h)/(\nu h),$$

$$\text{cond}(Z(V),D,h(s)) = (E(Z(V),D,h(s))/\nu)/h(s) = E(Z(V),D,h(s))/(\nu h(s)),$$

$$\text{cond}(A,D,h(s)) = (E(A,D,h(s))/\nu)/h(s) = E(A,D,h(s))/(\nu h(s)),$$

$$\text{cond}^*(Z(V),D) = \lim_{h \to 0} \text{cond}(Z(V),D,h),$$

$$\text{cond}(Z(V),D) = \lim_{s \to \infty} \text{cond}(Z(V),D,h(s)),$$

$$\text{cond}(A,D) = \lim_{s \to \infty} \text{cond}(A,D,h(s)).$$

Recall that $h(s) = 2^{1-s}$ is equal to the unit chopping error, that is, to the maximum relative input error caused by chopping to s bits, compare Notation 23.1. Therefore

$$\left.\begin{array}{c} \mathrm{cond}(Z(V),D,h(s)) \leq \mathrm{cond}(Z(V),D,h) \text{ if } h=h(s), \\ \mathrm{cond}(Z(V),D)) \leq \mathrm{cond}^{*}(Z(V),D)). \end{array}\right\} \quad (25.4)$$

The above definition of the conditions in a domain D is not efficient in the cases where the outputs Z(V) of the computational problem very greatly vary in D, as is the case for the problems of the evaluation of the product $v_0 v_1 \ldots v_{n-1}$ in the domain $\{|v_i| < \mu$ for all $i\}$ and of the power v^n in the domain $|v| < \mu$ for $\mu > 1$. (In such cases we may use Definition 25.2 provided that (25.1) holds.) The variation of the outputs in the cases of the problems MM, SLE, MI, is not great, however; see Section 29 about the case of Det.

Next we will estimate the condition of the problem of MM provided that the ℓ_∞-norm has been used, compare (24.5). We will recall (24.9) and will immediately deduce that

$$||\Delta(XY)|| \leq ||\Delta X||*||Y|| + ||\Delta Y||*||X||, \quad (25.5)$$

compare Proposition 26.3 in the next section. Let at first the input matrix Y be fixed so that $\Delta Y = 0$, and let the input matrix X range in the domain $D = D(\mu,\nu)$ such that

$$||X|| \leq \mu, \ \nu = \mu \, ||Y||.$$

Then it follows from (25.5) that $E(XY,D,h)/\nu \leq h$ and therefore

$$\mathrm{cond}(XY,D,h) \leq 1 \text{ for all positive } \mu,\nu \text{ and } h \text{ and for } D = D(\mu,\nu).$$

Similarly $\mathrm{cond}(XY,D,h) \leq 1$ if the input matrix X is fixed and the matrix Y is perturbed. Finally consider the case where both matrices X and Y may range in a domain D and both of them may be perturbed. Let the ℓ_∞-norm be applied again. Let the domain $D = D(\mu,\nu)$ be defined by the relations

$$||X|| + ||Y|| \leq \mu, \ ||XY|| = \nu = \mu^2/4. \quad (25.6)$$

(Here the bound on $||XY||$ follows from the bound on $||X||+||Y||$ and from (24.3).) Recall (25.3) and deduce that $E(XY,D,h) = \text{maximum } ||\Delta(XY)||$ where the maximum is in all $X, Y, \Delta X, \Delta Y$, such that (25.6) holds and

$$||\Delta X|| + ||\Delta Y|| \leq h\,\mu. \quad (25.7)$$

Combining (25.5)-(25.7) we deduce that

$$E(XY,D,h) \le (||\Delta X|| + ||\Delta Y||) \text{ maximum } \{||X||,||Y||\} \le h\,\mu^2.$$

Therefore (see (25.6)), cond(XY,D,h) $\le$ 4.

Taking into account also (25.4), we arrive at the following result, which shows that the MM problem is very well-conditioned.

Theorem 25.1. For all positive μ', μ'', h and for all natural s,

$$\text{cond}(XY,D,h) \le 4,\ \text{cond}(XY,D,h(s)) \le 4,\ \text{cond}(XY,D) \le \text{cond}^*(XY,D) \le 4$$

where D is defined by (25.6).

Finally we note that we may immediately derive some upper bounds on the bit-time, the bit-space, and the condition of an algorithm A if we know E(A,D,h(s)). Particularly we know E(A,D,h(s)) in the case of algorithms A=BA(n) and λ-algorithms A=BA(n,λ) for nXn MM, see Theorem 24.2. Therefore

$$\text{cond}(A,D,h(s)) \le M(M+2n^2+1)n^5 f\,f'f'' + O(h(s)) \tag{25.8}$$

in the case where A is a bilinear algorithm BA(n) of rank M for nXn MM applied in the domain (24.10) and where f, f', f'' are defined by (23.4). (25.8) means that chopping throughout the computation to $s + \lceil \log(M(M+2n^2+1)n^5 f\,f'f'') \rceil$ binary bits results in the output errors that are not greater than ones caused by chopping only the outputs to s bits. Thus the algorithm is well-conditioned as $s \to \infty$.

In the sequel we will derive some upper estimates for E(A,D,h(s)) in the case of the problems SLE(n), MI(n) and Det(n), and this will immediately imply some upper bounds on both bit-complexity and condition of those problems.

We will conclude this section with the analysis of λ-algorithms for MM. Theorem 23.7 gives only the trivial bound cond(A,D) $\le \infty$ if A is a bilinear λ-algorithm BA(n,λ) of positive degree d defined in the domain (24.10). In that case, the following quantities carry more information about the algorithm than cond(A,D) does.

$$\left.\begin{aligned} \text{logcond}(A,D,h(s)) &= \log(\nu/E(A,D,h(s)))/(s-1), \\ \text{logcond}(A,D) &= \lim_{s \to \infty} \text{logcond}(A,D,h(s)). \end{aligned}\right\} \tag{25.9}$$

We will call the latter quantities the s- <u>bit logarithmic condition of the algorithm</u> A <u>in the domain</u> D <u>and the asymptotic logarithmic condition of</u> A <u>in</u> D, respectively. (The following rule characterizes the meaning of the logarithmic conditions: <u>if chopping the floating point fractions of the outputs</u> Z(V) <u>to</u> s <u>binary bits</u>

introduces the error whose norm is bounded by E in a given domain D, then it is sufficient to keep only s logcond(A,D,h(s)) binary bits in the floating point fractions of all inputs and of all values computed by the algorithm A in order to assure that the output error norm does not exceed E in the domain D.

Theorems 23.7 and 24.2 immediately imply that

$$\text{logcond}(BA(n,\lambda),D) \leq (d+1)(1+q) \qquad (25.10)$$

provided that d is the degree of $BA(n,\lambda)$, λ is defined by (23.10) for $e(n)=E(A,D,h(s))$, and that

$$\left.\begin{aligned} q &= \lim_{n\to\infty} \log\,((d+1)^3M^2n^4\tilde{f}\,\tilde{f}'\tilde{f}'')/\log(\mu'(n)\mu''(n)/e(n)); \\ q&=0 \text{ if } \log(\mu'(n)\mu''(n)/e(n))/\log\,(dn) \to 0 \text{ as } n \to \infty . \end{aligned}\right\} \qquad (25.11)$$

If, say, q=0, d=1, and n is large, then the bound (25.9)-(25.10) imply that the precision of the λ-algorithm $BA(n,\lambda)$ need to be at most double comparing with the precision of the output, see the above rule for the logarithmic condition.

26. Estimating the Errors of the QR-factorization of a Matrix.

Our next objective is to estimate the values E(A,D,h(s)) for some of our algorithms A for SLE(n), MI(n), and Det(n). This will give us some upper bounds on the bit-time, the bit-space, and the condition of those algorithms, see Proposition 23.1 and Definition 25.3. If we apply the elimination processes such as ones defined by (19.5) and (19.6), we will arrive at the problem of preventing the magnification of errors. Pivoting is a customary means in that situation, see [A], p. 444; [Ric], pp. 134-135, but such a means becomes too costly in our case where we require $o(N^3)$ time and operate with block-submatrices rather than with numbers. Thus we will use QR-factorization to avoid the magnification of errors. In this section we will estimate the errors of Algorithm 20.2 applied with chopping to s bits.

In this part of the monograph, we will frequently refer to the unit relative chopping error $h=h(s)=2^{1-s}$, see Notation 23.1.

The following assumptions will simplify our analysis and notation. (Those assumptions will be used throughout the remainder of this part of the monograph.)

Assumption 26.1. All estimates are expressed in the ℓ_2-norm (unless it is explicitely stated otherwise) and the notation $||W||$ and $||\underline{v}||$ is used for $||W||_2$ and $||\underline{v}||_2$.

Assumption 26.1 will enable us to apply the relations (24.8).

Assumption 26.2. All inputs are real, compare Remark 23.3.

Assumption 26.3. The smaller terms of the higher order in $h=h(s)=2^{1-s}$ are ignored in our estimates, that is, we represent every polynomial in h by its term of the lowest degree in h. (We assume that $h \to 0$, $s \to \infty$, compare Remark 26.3 below.)

Remark 26.1. As this can be easily checked, the above deletion of the terms of the higher orders in h changes our resulting estimates for the bit-time, the bit-space, and the condition of our algorithms at most by the factor $1+O(h)$ as $h \to 0$ provided that h decreases sufficiently fast, see Remark 26.3 below.

Assumption 26.4. n is a power of 2, compare Remark 20.1.

We will need the following auxiliary results.

Proposition 26.1. Let X and Q be two mXn matrices, R be an nXn matrix such that

$$X = QR, \ Q^H Q = I. \tag{26.1}$$

Then

$$||X|| = ||R||. \tag{26.2}$$

Proof. Apply (24.3),(24.8), and (26.1) and derive that

$$||X|| = ||QR|| \le ||Q||^* ||R|| = ||R||,$$

$$||R|| = ||Q^H QR|| = ||Q^H X|| \le ||Q^H||^* ||X|| = ||X||.$$

This immediately implies (26.2). ■

Recall the definition of the ℓ_2-norm (see Example 24.1 and compare (24.6)), the notation $\Delta(W,s)$ (see Definition 25.1), and the bound 2^{1-s} on the relative error of chopping (see Notation 23.1) and deduce the following estimate.

Proposition 26.2. For arbitrary real number u and mXn matrix W,

$$|\Delta(u,s)| \le h|u|, \ ||\Delta(W,s)|| \le h\, m^{1/2}\, ||W(s)|| \text{ where } h=h(s)=2^{1-s}.$$

Furthermore there exist numbers u and mXn matrices W such that the equalities hold in the above.

Proposition 23.1 and the following simple results will be used in order to estimate the error propagation caused by arithmetical operations.

Proposition 26.3. For all matrices U, V, ΔU, and ΔV, which in particular may represent vectors or numbers, and for all real numbers c,

$$||(U + \Delta U) \pm (V + \Delta V) - (U \pm V)|| \leq ||\Delta U|| + ||\Delta V||,$$

$$||(U + \Delta U)(V + \Delta V) - UV|| \leq ||U||*||\Delta V|| + ||\Delta U||*||V|| +$$

$$||\Delta U||*||\Delta V||,$$

$$||(c + \Delta c)(U + \Delta U) - cU|| \leq |c|*||\Delta U|| + |\Delta c|*||U|| + |\Delta c|*||\Delta U||.$$

Proposition 26.4, see [A], p. 425. Let U and ΔU be two nXn matrices such that U is invertible and $||\Delta U|| < 1/||U^{-1}||$. Then the matrix $U + \Delta U$ is invertible and

$$||(U + \Delta U)^{-1}|| \leq ||U^{-1}||/(1-||U^{-1}||*||\Delta U||).$$

Proposition 26.5, see [A], p. 463. Let the assumptions of Proposition 26.4 hold. Let in addition V, Z, ΔV, and ΔZ be four nXp matrices such that UZ=V and $(U + \Delta U)(Z + \Delta Z) = V + \Delta V$. Then

$$||\Delta Z|| \leq ||U^{-1}||(||\Delta V|| + ||U^{-1}||*||V||*||\Delta U||)/(1-||U^{-1}||*||\Delta U||).$$

Here p and n can be arbitrary positive integers.

Hereafter Q(s) and R(s) designate the approximations to the QR-factors of a matrix X computed by Algorithm 20.2 with chopping to s bits throughout the computation, compare Definition 25.1.

In this section we will show that $Q^H(s)$ is a near unitary matrix and we will estimate $||Q^H(s)X - R(s)||$ in the domains D where

$$||X|| \leq \mu \tag{26.3}$$

for a finite $\mu = \mu(n)$. ((26.2) implies that $||R|| \leq \mu$ in D. We will apply our estimates to the analysis of the bit-complexity and of the condition of the problems SLE, MI, and Det. The bounds on $||Q(s)-Q||$ and $||R(s)-R||$ will not be needed in our analysis of SLE,MI, and Det and will not be derived.)

Algorithm 20.2 solves Problem 20.1 in the special case where $\tilde{R} = 0_n$. In that case the inputs and outputs of the algorithm are just the nXn matrix X and the pair of nXn matrices R and Q, respectively. This enables us to save the notation $\tilde{Q}$ for the 2X2 matrix in (20.9),

$$\tilde{Q} = (1/r)\begin{bmatrix} x^H & \tilde{r}^H \\ -\tilde{r} & x \end{bmatrix}, \quad r = (x^H x + \tilde{r}^H\tilde{r})^{1/2}.$$

Let us consider the steps (20.9) where we evaluate $\tilde{Q}$ and r from given inputs x and $\tilde{r}$ using chopping to s bits throughout the computation and let x(s), $\tilde{r}(s)$, $\tilde{Q}(s)$, and r(s) designate the computed approximations to x, $\tilde{r}$, $\tilde{Q}$, and r, respectively, compare Definition 25.1. Hereafter $\underline{v}^T$ will designate the <u>transpose</u> of a vector <u>v</u> (actually $\underline{v}^T = \underline{v}^H$ for real vectors <u>v</u> and similarly u^H=u for real numbers u but here we will not refer to Assumption 26.2 and will use the superscripts T and H as if we dealt with nonreal values.) We will designate

$$[\Delta'(s), \Delta''(s)]^T = \tilde{Q}(s)\,[x(s),\tilde{r}(s)]^T - [r(s),0]^T. \tag{26.4}$$

Then we will have the following estimates.

<u>Proposition 26.6</u>. $|\Delta(r,s)| = |r(s)-r| \le 2hr$; $\tilde{Q}(s)\tilde{Q}^H(s) = \tilde{Q}^H(s)\,\tilde{Q}(s) = (1+\Delta(s))I$ for a constant $\Delta(s)$ such that $|\Delta(s)| \le 6h$; $|\Delta'(s)| \le 3hr$; $|\Delta''(s)| \le hr$. Here $\Delta'(s)$ and $\Delta''(s)$ are defined by (26.4).

<u>Proof</u>. It follows from (20.9) that

$$\tilde{Q}(s) = \begin{bmatrix} (x/r)_s^H & (\tilde{r}/r)_s^H \\ -(\tilde{r}/r)_s & (x/r)_s \end{bmatrix}$$

where $(x/r)_s$ and $(\tilde{r}/r)_s$ designate the computed approximations to x/r and $\tilde{r}/r$, respectively. This immediately implies that the matrix equation of Proposition 26.6 holds for some real $\Delta(s)$. It remains to estimate $|\Delta(r,s)|$, $|\Delta(s)|$, $|\Delta'(s)|$, and $|\Delta''(s))|$ from above.

To simplify the notation, assume that x and $\tilde{r}$ themselves have s-bit floating point fractions, so that x=x(s), $\tilde{r}=\tilde{r}(s)$. (This assumption does not influence the resulting estimates.) We have the following bound on the error of the computed approximation to r^2 (ignoring the terms $O(h^2)$, see Assumption 26.3),

$$|\Delta(r^2,s)| = |\Delta(x^H x + \tilde{r}^H\tilde{r},s)| \le 2h(x^H x + \tilde{r}^H\tilde{r}) = 2hr^2,$$

see Proposition 23.2. This implies the desired estimate

$$|\Delta(r,s)| = |\Delta((x^H x + \tilde{r}^H\tilde{r})^{1/2},s)| \le 2hr.$$

(Indeed the relative error decreases by one half when the square root has been evaluated but then increases by at most h due to chopping, see Proposition 26.2.)

Similarly the divisions of x and $\tilde{r}$ by r preserve the same relative error (ignoring the errors of higher orders) but chopping adds h to the upper bound 2h on the relative error, see Proposition 26.2, so that we have

$$|\Delta(x/r,s)| \leq 3h|x/r|, \quad |\Delta(\tilde{r}/r,s)| \leq 3h|\tilde{r}/r|.$$

Combining the latter bound with Proposition 26.3 we obtain the desired bound on $|\Delta(s)|$.

Apply Proposition 26.3 and deduce that

$$|\Delta''(s)| \leq |\Delta(\tilde{r}/r,s)|*|x| + |\Delta(x/r,s)|*|\tilde{r}|.$$

Here we may just substitute the above bounds on $|\Delta(x/r,s)|$ and $|\Delta(\tilde{r}/r,s)|$ but we will get a little better bound noting that $-(\tilde{r}/r)x + (x/r)\tilde{r} = 0$ so that the error of the evaluation of r can be ignored. Then we need estimate only the errors $\Delta(x/r,s)$ and $\Delta(\tilde{r}/r,s)$, caused by the chopping of the outputs of the divisions by r, and we obtain that $|\Delta''(s)| \leq 2h|x\tilde{r}/r|$ (within the errors of the order $O(h^2)$). This implies the desired bound on $|\Delta''(s)|$ because $2|x\tilde{r}| \leq |x|^2 + |\tilde{r}|^2 = r^2$.

Similarly we derive that

$$\Delta'(s) = (x/r(s))(1+h_1)x+(\tilde{r}/r(s)(1+h_2)\tilde{r}-r(s) = (1/r(s))(x^2+\tilde{r}^2-r^2(s)+h_1x^2+h_2r^2)$$

where r(s) is the computed approximation to r; h_1 and h_2 are the relative errors of the divisions by r(s) performed with chopping to s bits, $|h_1| \leq h$, $|h_2| \leq h$. Therefore $|\Delta'(s)| \leq h\, r^2/r(s)+|(r^2-r^2(s)|/r(s)$. Substitute the already available bound $|r^2-r^2(s)| = |\Delta(r^2,s)| \leq 2hr^2$ and deduce that $|\Delta'(s)| \leq 3hr^2/r(s)$. Since $r/r(s)=1+O(h)$ we derive the desired bound on $|\Delta'(s)|$ ignoring the terms $O(h^2$.) ■

<u>Notation</u> <u>26.1</u>. Hereafter in this section f, f', f'' will designate the maximum magnitudes of the constants involved in the MM algorithms included in Algorithm 20.2, compare (23.4). M will bound the rank of those MM algorithms. We will assume that $M \leq n^3$ and that $f=O(n)$, $f'=O(n)$, $f''=O(n)$ as $n \to \infty$, compare (2.12) and the proof of Corollary 23.4.

Proposition 26.6 can be applied to all steps (20.9) of Algorithm 20.2 where the 2X2 unitary submatrices Q_s of the nXn unitary matrices $\tilde{Q}_s$ are evaluated for $s=1,2,\ldots,n^2$, see (20.9). (Actually several matrices Q_s turn into I_2 and this observation could be used in order to improve our estimates.) All other steps of Algorithm 20.2 are just multiplications by unitary matrices, which cannot increase the norms of the errors, because all unitary matrices have norm 1, see (24.3) and (24.8). This leads us to the next result.

<u>Proposition</u> <u>26.7</u>. Let $\tilde{Q}_k(s)$, $k=1,2,\ldots,n^2$, designate the approximations to the nXn matrices $\tilde{Q}_k$ in (20.2) computed by Algorithm 20.2. Let

$Q^*(s) = \tilde{Q}_{n^2}(s)\tilde{Q}_{n^2-1}(s)\ldots\tilde{Q}_1(s)$. Then (under Notation 26.1) $||Q^*(s) - Q^H(s)|| \leq hn^7M(M+2n^2+1))\ f\ f'f = O(hn^{16})$ as $n \to \infty$, $h \to 0$.

Proof. The matrix $Q^H(s)$ has been computed by multiplying the n^2 matrices $\tilde{Q}_k(s)$, $k=1,2,\ldots,n^2$, with each other. The norm of every matrix $\tilde{Q}_k(s)$ is bounded by 1+3h by the virtue of Proposition 26.6. Consequently the multiplication by the products of n^2 such matrices may magnify the errors only negligibly (at most by the factor $1+3hn^2$). Therefore the norm of the total error is bounded by the sum of the norms of the error matrices of all n^2 MM's. n^2 successive applications of Theorems 23.3 and 24.2 (where $\mu'(n) = \mu''(n) = 1, M \leq n^3$) yield the inequality of Proposition 26.7 (which is a rather rough bound because each $\tilde{Q}_k(s)$ differs from I_n only within a 2X2 submatrix). ■

For the estimates in the next two sections we need have a unitary matrix Q^* near $Q^H(s)$ such that $\det Q^* = 1$. That objective will be achieved by combining Proposition 26.7 with the next result.

Proposition 26.8. There exists a nXn unitary matrix Q^* such that $\det Q^* = 1$ and $||Q^*(s) - Q^*|| \leq 3hn^2 + O(h^2)$.

Proof. As follows from Proposition 26.6, there exist positive constants $g_k(s)$ such that

$$|g_k(s) - 1| \leq 3h$$

and

$$(g_k(s))^2 Q_k^H(q) Q_k(s) = I_2 \text{ for } k=1,2,\ldots,n^2.$$

Then the 2X2 matrices $Q_k^*(s) = g_k(s)Q_k(s)$ are unitary. Substitute them for $Q_k(s)$ while defining the nXn matrices $\tilde{Q}_k(s)$. This will transform the nXn matrices $\tilde{Q}_k(s)$ into nXn unitary matrices $\tilde{Q}_k^*(s)$. Choose

$$Q^* = \tilde{Q}^*_{n^2}(s)\tilde{Q}^*_{n^2-1}(s)\ldots\tilde{Q}^*_1(s),$$

The matrix Q^* is unitary since it is the product of unitary matrices.

We apply the version of Algorithm 20.2 with the normalization; therefore $\det Q^* = 1$, compare the end of Section 21.1. The inequality of Proposition 26.8 follows because, ignoring the terms $O(h^2)$, we have

$$||Q^*(s) - Q^*|| \leq \sum_{k=1}^{n^2} ||Q_k^*(s) - Q_k(s)|| \leq 3hn^2. \ ■$$

Corollary 26.9. There exists an nXn unitary matrix Q^* such that $\det Q^* = 1$ and $||Q^* - Q^H(s)|| \leq hn^2(M(M+2n^2+1)n^5 f\ f'f'' + 3) = O(hn^{16})$ as $n \to \infty$, $h \to 0$.

Similarly we will estimate $||Q^H(s)\ X - R(s)||$. Algorithm 20.2 can be reduced to the n^2 successive premultiplications of X by the unitary matrices $\tilde{Q}_k, k=1,2,\ldots,n^2$,

$$X_k = \tilde{Q}_k X_{k-1},\ k=1,\ldots,n^2,\ X_0 = X,\ X_{n^2} = R.$$

(The actual order of performing MM's in Algorithm 20.2 is different, that is, some $\tilde{Q}_k$ are first multiplied together and then their product premultiplies X_i. This will not change our resulting estimates, however.) Computing with chopping to s bits we obtain that

$$\left.\begin{aligned} &X_k(s) = \tilde{Q}_k(s)\ X_{k-1}(s) + \Delta_k(s) + \Delta_k''(s), \\ &k=0,1,\ldots,n^2,\ X_0(s) = X(s),\ R(s) = X_{n^2}(s). \end{aligned}\right\} \quad (26.5)$$

Here $X_k(s)$, $\tilde{Q}_k(s)$ denote the computed approximation to X_k, $\tilde{Q}_k$, respectively; $\Delta_k(s)$ is the matrix of the errors caused by the premultiplication of $X_{k-1}(s)$ by $Q_k(s)$, $\Delta_k''(s)$ is the matrix of the errors caused by the substitution of 0 for an appropriate subdiagonal entry of $\tilde{Q}_k(s)X_{k-1}(q)$. ($\Delta_k''(s)$ corresponds to $\Delta''(s)$ of (26.4).) Recall that all subdiagonal entries must finally turn into 0 *to make the output matrix* R(s) *triangular*.) As follows from Proposition 26.6,

$$||\Delta_k''(s)|| \leq h\ \mu. \quad (26.6)$$

$\Delta_k(s)$ is the error of pXp MM for $p \leq n$, so we may apply Theorems 23.3 and 24.2 where $\mu'(n)=1$, $\mu''(n) = \mu$ and derive that

$$||\Delta_k(s)|| \leq hn^5 M(M+2n^2+1)\ f\ f' f''\ \mu. \quad (26.7)$$

The subsequent premultiplications of $X_k(s)$ by $Q_{k+j}(s)$ for $j=1,2,\ldots,n^2-k$ may magnify the norm of the error matrix only negligibly because the matrices $Q_{k+j}(s)$ are near unitary and consequently have the norms near 1, see Proposition 26.7 and the relations (24.3),(24.8). Therefore we just sum the bounds (26.6) and (26.7) on $||\Delta_k(s)||$ and $||\Delta_k''(s)||$ in all k, $k=1,2,\ldots,n^2$, in order to obtain that

$$||Q^*(s)\ X - R(s)|| \leq h\ \mu\ n^7 M(M+2n^2+2) f\ f' f'' = O(h\ \mu\ n^{16}) \text{ as } n \to \infty,\ h \to \infty.$$

Here the values M, f, f', f'' are defined by the algorithms employed for MM's within Algorithm 20.2.

Combining the latter bound with Proposition 26.7, we derive the following result.

Theorem 26.10. Let Q(s) and R(s) be the approximations to the QR-factors of an nXn matrix X computed by Algorithm 20.2 with chopping to s bits throughout. Then

$$||Q^H(s)\ X - R(s)|| \leq 2h\ \mu\ n^7\ M(M+2n^2+2)f\ f'f'' = O(h\ \mu\ n^{16}) \qquad (26.8)$$

as $n \to \infty$, $h \to 0$, see Notation 26.1.

Remark 26.2. It can be assumed in Corollary 26.9 and in Theorem 26.10 that $M = O(n^{\omega+\varepsilon}) = o(n^{2.5})$, compare Theorem 17.1. Then we deduce that

$$||Q^* - Q^H(s)|| = o(hn^{15}), \qquad (26.9)$$

$$||Q^H(s)X - R(s)|| = o(hn^{15}\mu) \qquad (26.10)$$

provided that $n \to \infty$.

Remark 26.3. To justify the application of Assumption 26.3 in this section it suffices to require that

$$\log(n\ \mu\ )/\log(1/h) \to 0 \text{ as } n \to \infty\,. \qquad (26.11)$$

In the subsequent similar estimates we will require that h rapidly converge to zero also in comparison with some other parameters, μ',ν,E. Generally it will suffice to require that

$$\log(\ \mu\ \mu'\ \nu\ n/E)/\log(1/h) \to 0 \qquad (26.12)$$

substituting 1 where (some of the values) E,μ,μ',ν are undefined.

In the next sections we will derive some asymptotic estimates refering to (26.9)-(26.12). For the extension to the nonasymptotic estimates, apply Corollary 26.9 and Theorem 26.10 at the place of (26.9) and (26.10), respectively.

27. The Bit-Complexity and the Condition of the Problem of Solving a System of Linear Equations.

In this section we will estimate the precision E(A,D,h(s)) and consequently the bit-time, the bit-space, and the condition of the algorithm A from Section 21 for SLE(n), that is, for the evaluation of the n-dimensional vector $\underline{z} = X^{-1}\underline{v}$, compare (21.1). Here we assume that an nXn invertible matrix X and an n-dimensional vector $\underline{v}$ range in the domain D defined by the following inequalities, compare (26.4),

$$||X|| \leq \mu,\ ||X^{-1}|| \leq \nu,\ \mu\ \nu \geq 1, \qquad (27.1)$$

$$||\underline{v}|| \leq \mu', \qquad (27.2)$$

and by some positive constants $\mu = \mu(n)$, $\nu = \nu(n)$, and $\mu' = \mu'(n)$. We require that $\mu\ \nu \geq 1$ since $||X||*||X^{-1}|| \geq 1$, see (24.3),(24.7). (Without bounding $||X^{-1}||$ we may run into unbounded outputs whose approximation with a prescribed positive absolute precision requires unbounded bit-time and bit-space.)

We will assume that the algorithm A includes Algorithm 20.2 at the stage of the evaluation of the QR-factors of X and that A performs with chopping to s bits throughout. We will continue using Assumptions 26.1-26.4 and the notation of the previous section. Then $\underline{z}(s)$, the output approximation to $\underline{z} = X^{-1}\underline{v}$, will be computed by back substitution applied to the following triangular system of linear equations in $\tilde{\underline{z}}(s)$,

$$R(s)\ \tilde{\underline{z}}(s) = \underline{u}(s), \tag{27.3}$$

so that $\tilde{\underline{z}}(s) = (R(s))^{-1}\underline{u}(s)$. Here $\underline{u}(s)$ is the computed approximation to the vector $Q^H(s)\underline{v}$; Q(s) and R(s) denote the computed approximations to the QR-factors of X.

Recall Proposition 23.2 and immediately deduce that

$$||\underline{u}(s) - Q^H(s)\underline{v}|| \leq hn^2\mu'. \tag{27.4}$$

Next apply backward error analysis, see [Wil], pp. 209-215, or [CdB], p. 181, and derive that

$$(R(s) + \Delta)\ \underline{z}(s) = \underline{u}(s). \tag{27.5}$$

Here Δ is the backward error matrix such that

$$|\Delta| \leq h\ |R(s)|; \tag{27.6}$$

the notation $|W|$ stands for the matrix $[|w_{ij}|]$ provided that $W = [w_{ij}]$.

The latter inequality implies that

$$||\Delta||_\infty \leq h\ ||R(s)||_\infty.$$

Since we prefer to work with the ℓ_2-norm and do not care much about the factor $n^{1/2}$ in our estimates, we will slightly worsen the latter bound and will deduce from (27.6) and from the definition of the ℓ_2-norm, see Example 24.1, that

$$||\Delta|| \leq hn^{1/2}\ ||R(s)|| \leq h\ \mu\ n^{1/2}. \tag{27.7}$$

((26.9) and (27.1) imply that $||R|| \leq \mu$. Applying (26.8) and Corollary 26. 9 we extend that bound to $||R(s)|| \leq \mu + O(h)$. By Assumption 26.3, we may ignore the term $hn^{1/2}O(h)$ and we write $||R(s)|| \leq \mu$.) Combine (26.10) and (27.7) and derive that

$$||R(s) + \Delta - Q^H(s)X|| = o(hn^{15}\mu) \text{ as } n \to \infty. \tag{27.8}$$

Next compare (27.5) with the matrix-vector equation

$$Q^H(s)\ X\ \underline{z} = Q^H(s)\underline{v} \tag{27.9}$$

(which is equivalent to (21.1), that is, has the same solution $\underline{z} = X^{-1}\underline{v}$). Apply Proposition 26.5 for $p=1$, $Z=\underline{z}$, $\Delta Z=\underline{z}(s)-\underline{z}$, $V=Q^H(s)\underline{v}$, $\Delta V=\underline{u}(s)-Q^H(s)\underline{v}$, $U=Q^H(s)X$, $\Delta U=R(s)+\Delta - Q^H(s)X$; also apply Corollary 26.9 (see (26.9)); Assumption 26.3; the inequalities (27.1),(27.2),(27.4), and (27.8) and derive the following estimate for $E(A,D,h(s)) = ||\underline{z}(s)-\underline{z}||$.

<u>Theorem 27.1</u>. Let A be the algorithm from Section 21 for the evaluation of the vector $\underline{z} = X^{-1}\underline{v}$ defined in a domain D where (27.1) and (27.2) hold. Let the auxiliary QR-factors of X be computed by Algorithm 20.2. Let $n \to \infty$ and $s \to \infty$ and let (26.12) hold. Then, under Assumptions 26.1-26.4,

$$E(A,D,h(s)) = o\ (hn^{15}\mu'\ \mu\ \nu^2) \le chn^{15}\mu\ \mu'\ \nu^2 \tag{27.10}$$

for a positive constant c.

Combining Theorems 22.1 and 27.1 with Proposition 23.1, see also Remark 22.2, we derive the following corollary.

<u>Corollary 27.2</u>. Under the assumptions of Theorem 27.1, let $E = E(A,D,h(s))$ satisfy (27.10); let $bt(A,D,E)$, $bs(A,D,E)$ denote the bit-time and the bit-space of the algorithm A, compare the end of Section 23; let ω,ε, and $bt(s)$ be defined as in Theorem 23.6. Then simultaneously

$$bt(A,D,E) = O(n^{\omega+\varepsilon}\ bt(s)),\ bs(A,D,E) = O(n^2 s) \tag{27.11}$$

for $s=o(\log(n^{15}\mu'\ \mu\ \nu^2/E))$ as $s \to \infty$ and $n \to \infty$.

<u>Remark 27.1</u>. Let us show that $\mu'\ \nu/E$ is a lower bound on $h(s)$ and that we have to use at least $n^2 s = n^2(1+\log(\mu'\ \nu/E))$ binary bits in the floating point fractions of the entries of X in order to have the norm of the output error $\Delta\underline{z}$ bounded by E, that is, to have $||\Delta\underline{z}|| \le E$. Indeed, we have the output error $\Delta\underline{z} = (X^{-1}(s) - X^{-1})\underline{v} = -X^{-1}(s)\ \Delta\ (X,s)X^{-1}\underline{v}$. For fixed s, μ', and ν, we will choose $X(s)$, X, and $\underline{v}$ such that

$$||\Delta\underline{z}|| = ||X^{-1}(s)\ \Delta\ (X,s)X^{-1}\underline{v}|| \ge h(s)\ \mu'\nu + O(h^2(s)).$$

To satisfy that bound, choose X,s, and ν such that

$$X = \alpha\ I,\ \alpha = (1/\ \nu) + O(h(s)),\ ||\ \Delta\ (X,s)|| = h(s)||X||,$$

compare Proposition 26.2. Then, choosing appropriate vector $\underline{v}$ and ignoring the values $O(h^2(s))$, we deduce that

$$||\Delta\underline{z}|| = h(s)\ \mu'\ \nu,\ s-1 = \log(\ \mu'\ \nu\ /||\ \Delta\underline{z}||).$$

Finally let $E = h(s)\mu'\nu$, then allow s vary, and note that $||\Delta \underline{z}|| \leq E$ would require that at least $s = \log(\mu'\nu/E))+1$ binary bits be used in the fraction of a diagonal entry of X. Similar argument extends that conclusion also to other entries of X. (Remark 24.1 can be extended to the case of SLE.) ∎

Finally, we will apply Theorem 27.1 in order to estimate cond(A,D), see Definition 25.3. (Note that in our case the norm of the output vector $\underline{z} = X^{-1}\underline{v}$ does not exceed the product $\nu\mu'$. Such a product should substitute for ν in Definition 25.3.)

Corollary 27.3. Under the assumptions of Theorem 27.1,

$$\text{cond}(A,D) = o(n^{15}\mu\nu) \text{ as } n \to \infty .$$

Remark 27.2. Compare the customary value $||X||*||X^{-1}||$ of the condition of the matrix X of the system of linear equations (21.1). Such a value is not less than 1 and is bounded from above by $\mu\nu$ in the domain (27.1),(27.2). Also Proposition 26.5 implies that $\text{cond}(X^{-1}\underline{v},D) \leq \mu\nu$. The factor $o(n^{15})$ is due to the roughness of our estimates and to the overhead of our algorithm.

28. The Bit-Complexity and the Condition of the Problem of Matrix Inversion.

In this section we will extend our analysis to the case of the problem MI(n). We will keep computing the QR-factors of the given matrix X by Algorithm 20.2 and will keep using Assumptions 26.1-26.4 and Notation 23.1.

The algorithm A for MI(n) will be taken from Section 21, see (21.4)-(21.8). We will apply that algorithm in the domain D defined by (27.1). The precision E(A,D,h(s)) is estimated similarly to the previous section. In particular the estimates for the errors at the QR-factorization stage do not change.

The major difference with the previous section is that now we cannot afford back substitution, which would involve too many (at least n^3) arithmetical operations for TMI(n). Thus we will apply the process (21.4)-(21.8).

Let us estimate the errors omitting some minor details. (26.1) and (27.1) imply that

$$||R|| \leq \mu,\ ||R^{-1}|| \leq \nu. \tag{28.1}$$

(Indeed, compare Proposition 26.1, note that $R^{-1} = X^{-1}Q$, and apply (24.3) and (24.8). ∎)

The upper bounds (28.1) are extended to the norms of all submatrices of R used in the inversion algorithm, compare (21.6),(24.6) and Proposition 26.4. Therefore the errors caused by chopping of the entries of R and of any matrix computed in the process of the inversion of R may be magnified by at most 2 log n subsequent MM's, so that the total magnification is at most by the factor

$$(M(M+2n^2+1)n^5 f\, f' f'' \mu\, \nu)^{2 \log n} = O(n^{\alpha} \mu\, \nu)^{2 \log n}, \tag{28.2}$$

see Theorems 23.3 and 24.2. Here M is the maximum rank of the MM's; f, f', f'' are the maximum magnitudes of the constants of those MM's, compare (23.4) and the proof of Corollary 23.4;

$$\alpha = \alpha(M) = 14 \text{ if } M \le n^3,\ \alpha = \alpha(M) < 13 \text{ if } \log M/\log n < 2.5.$$

Apart from MM's, also n inversions of the diagonal entries of R are involved at the stage of the inversion of R. The outputs of those n inversions are the diagonal entries of R^{-1} so that their magnitudes are at most ν, see (24.6) and (28.1). The relative and absolute errors of those n inversions are at most $h=h(s)$ and at most $h\,\nu$, respectively. Combining the latter estimate with (28.2) and with the bound $hn^{1/2}\mu$ on the absolute error of chopping the inputs (see Proposition 26.2), we arrive at the total bound on the norm of the absolute error of the inversion of R,

$$E(R^{-1}, D, h(s)) = O(n^{\alpha} \mu\, \nu)^{2 \log n}(\mu\, n^{1/2} + \nu)h(s). \tag{28.3}$$

We may discard that smaller error as well as the error caused by the multiplication $X^{-1} = Q^H R^{-1}$ because of the dominating influence of the errors of the QR-factorization of X magnified by the factor (28.2) (the estimates for the latter errors can be obtained similarly to the analysis in the previous section, see (27.10) for $\mu' = 1$). We finally arrive at the following results.

Theorem 28.1. Let A be the algorithm (21.4)-(21.8) from Section 21 for the evaluation of $Z = X^{-1}$ (the inverse of a given $n \times n$ matrix X defined in the domain D where 27.1 holds). Let the auxiliary QR-factors be computed by Algorithm 20.2. Let $n \to \infty$ and (26.12) hold. Then, under Assumptions 26.1-26.4,

$$E(A, D, h(s)) = o(hn^2\nu)(n^{13}\mu\, \nu)^{2 \log n+1}. \tag{28.4}$$

Corollary 28.2. Under the assumptions of Theorem 28.1, let $E=E(A,D,h(s))$ satisfy (28.4); let bt(A,D,E) and bs(A,D,E) be defined in the end of Section 23; let ω, ε, bt(s) be defined as in Theorem 23.6. Then (27.11) holds where

$$s = O(\log n \log(n^{13}\, \mu\, \nu\,)^2 + \log(n^2\nu) - \log E). \tag{28.5}$$

Also

$$\text{cond}(A, D) = o(n^2(n^{13}\mu\, \nu)^{2 \log n+1}). \tag{28.6}$$

(28.6) does not show that our algorithm A is well-conditioned and actually it is rather ill-conditioned. Still (28.5) implies that the <u>exponent of the bit-time of MI remains</u> equal to ω, that is, to the exponent of MM, and that the bit-space of MI(n) is still of the order $(n \log n)^2$ provided that μ, ν, and E grow as polynomials in n.

<u>Remark 28.1</u>. Remark 27.1 can be extended to the case of MI(n) where $\mu' = 1$. Also we have the obvious lower bound $n^2\log(\nu/E)$ on the number of bits required in order to represent the output matrix X^{-1} with the precision E in the domain D.

<u>29. The Bit-Complexity and the Condition of the Problem of the Evaluation of the Determinant of a Matrix.</u>

In this section we will study the problem Det(n). Let the algorithm A for the evaluation of det X be defined by (20.2),(21.9),(21.10), and by the normalization equation $\det \tilde{Q}=1$. Let A be applied in a domain D where (26.3) holds. Then the computation is reduced to finding an approximation R(s) to one of the two QR-factors of X by Algorithm 20.2, that is, to the triangular matrix R. It remains to multiply the diagonal entries of R together.

Let us first estimate the precision of the first step, that is, let us estimate the value $|\det R(s) - \det X|$.

As follows from Corollary 26.9 and from Theorem 26.10, see also Remark 26.2, there exists a unitary matrix Q^* such that

$$||Q^*X - R(s)|| = o(hn^{15}\mu) \text{ as } n \to \infty \text{ and } h \to 0 \tag{29.1}$$

and in addition $\det Q^* = 1$. Consequently

$$\det X = \det(Q^*X). \tag{29.2}$$

We will compare det R(s) with $\det(Q^*X)$ taking into account (29.1) and (29.2).

Let hereafter $r_{ij}(s)$ and x^*_{ij} designate the entries (i,j) of the matrices R(s) and Q^*X, respectively, where both i and j range from 0 to n-1. We will use the following bounds.

<u>Proposition 29.1</u>. $|x^*_{ij}| \leq \mu$ for all i,j; $|x^*_{ij}| = o(hn^{15}\mu)$ if i>j; $|r_{ij}(s)| \leq \mu + o(hn^{15}\mu)$ for all i.

<u>Proof</u>. $|x^*_{ij}|$ equals the ℓ_2-norm of the 1X1 submatrix $[x^*_{ij}]$ of Q^*X and therefore $|x^*_{ij}| \leq ||Q^*X||$, see (24.6). This implies the first group of inequalities of Proposition 29.1 because $||Q^*X|| = ||X|| \leq \mu$, see (24.3),(24.8),(26.3). It remains to recall (29.1) and the fact that $r_{ij}(s)=0$ if i>j (since R(s) is a triangular matrix) in order to derive all other inequalities of Proposition 29.1.■

It is well known that for an arbitrary nXn matrix U, the value $|\det U|$ equals the n-dimensional volume of the parallelepiped P(U) in the n-dimensional Euclidean space, P(U) being defined by its edges, which are chosen to be the column-vectors of U, see [BML]. This implies the following fact.

Proposition 29.2. $|\det U| \le ||U||^n$ for all nXn matrices U and the equality holds in the above if and only if $U/||U||$ is a unitary matrix.

Let X_k^* designate the kXk submatrix located in the northwest (upper left) corner of Q^*X, k=1,2,...,n. Then decomposing $\det(Q^*X)$ along the last row of Q^*X and applying Propositions 29.1 and 29.2 we deduce that, for a certain constant c,

$$|\det (Q^*X) - x_{nn}^* \det X_{n-1}^*| \le chn^{15}(n-1)\ \mu^n.$$

Similarly

$$|x_{nn}^* \det X_{n-1}^* - x_{nn}^*\ x_{n-1,n-1} \det X_{n-2}^*| \le (chn^{15})^2(n-2)\ \mu^n,$$

and so on. Summarizing we derive that

$$|\det(Q^*X) - x_{00}^*x_{11}^*\cdots x_{n-1,n-1}^*| \le \sum_{i=1}^{n-1} (chn^{15})^i(n-i)\mu^n \le chn^{16}\mu^n(1-(chn^{15})^{n-1})/(1-chn^{15}).$$

Ignoring the smaller terms (see Assumption 26.3) we derive that

$$|\det (Q^*X) - x_{00}^*x_{11}^*\cdots x_{n-1,n-1}^*| \le chn^{16}\mu^n. \tag{29.3}$$

Next we will estimate the error $\Delta(\det R(s),s)$ of the evaluation of $|\det R(s)| = |r_{00}(s)r_{11}(s)\cdots r_{n-1,n-1}(s)|$ with chopping to s bits. Using backward error analysis (for instance, recursively using Proposition 26.2 for K=1) and ignoring the terms that are $O(h^2)$, see Assumption 26.3, we easily deduce that

$$|\Delta(\det R(s),s)| \le hn\ |\det R(s)|.$$

Since $|\det R(s)| \le \mu^n(1+O(h))$, see (29.1) and Propositions 29.1 and 29.2, we obtain (ignoring the terms $O(h^2)$) that

$$|\Delta(\det R(s),s)| \le hn\ \mu^n. \tag{29.4}$$

Combining (29.2)-(29.4) we derive the following desired bound on $E(A,D,h(s)) = |\Delta_A(\mathrm{Det},X,s)|$.

Theorem 29.3. Let the algorithm A from Section 21 for the evaluation of det X be defined by (20.2),(20.9),(21.10), and by the normalization equation $\det \tilde{Q} = 1$. Let A be applied in the domain D where (26.3) holds. Let the auxiliary evaluation of the QR-factors of X be performed by Algorithm 20.2. Then, under Assumptions 26.1-26.4, there exists a constant c such that

$$E(A,D,h(s)) \leq h(n + cn^{16})\, \mu^n. \tag{29.5}$$

Corollary 29.4. Under the assumptions of Theorem 29.3, let

$$s - 1 = \lceil \log((n + cn^{16})\, \mu^n/E) \rceil = \lceil n \log \mu + \log(n + cn^{16})) - \log E \rceil, \tag{29.6}$$

or equivalently, let (29.5) hold for $E=E(A,D,h(s))$. Let ω, ε, and bt(s) be as in Theorem 23.6. Let $n \to \infty$ and let (26.12) hold. Then (27.11) holds where s is defined by (29.6).

We can see a substantial increase of the upper bounds on the bit-time and bit-space comparing with the estimates of Corollaries 27.2 and 28.2, that is, roughly n times if all of μ, μ', and ν are the values of the same order of magnitude. Similarly the lower bounds increase.

Theorem 29.5.

$$bt(\det X,D,E) \geq n^2 s,\quad bs(\det X,D,E) \geq n^2 s$$

where the domain D is defined by (26.3) and $s - 1 = \lceil \log(\mu^n/E) \rceil$.

Proof. The error caused by chopping of the fraction of a single entry (i,j) of X to s bits may change its value by μh, see Proposition 26.2. Such a perturbation changes the output by $h \mu M(i,j)$ where $M(i,j)$ designates the value of the determinant of the $(n-1) \times (n-1)$ submatrix of X obtained by deleting row i and column j from X. $|M(i,j)|$ can be as large as μ^{n-1}, making the error caused by the perturbation of that entry of X as large as $h \mu^n$. (Examine, for example, the case where $i=j=1$ and $X=\mu I$.) To achieve the precision E, we should choose $h \leq E/\mu^n$, $s - 1 \geq \log(\mu^n/E)$. Since the latter argument can be applied to all entries of X, this implies Theorem 29.5. (Note that Remark 24.1 can be extended to the case of Det.) ■

Remark 29.1. In the above example where $X = \mu I$, chopping of the inputs causes small relative error of det X, that is, h or less. However, in the following example the relative error is as large as 1.

Example 29.1. Let

$$X = \begin{bmatrix} X_{11} & 0 \\ 0 & I \end{bmatrix}, \quad X_{11} = \begin{bmatrix} u & 1 \\ 1 & 1 \end{bmatrix}, \quad u = 1+2^{-r}.$$

Then $\det X = 2^{-r}$. If the input entries of X are chopped to s binary bits where $s \leq r$, then

$$X_{11}(s) = \begin{bmatrix} 1 & 1 \\ 1 & 1 \end{bmatrix},$$

$\det X(s) = \det X_{11}(s) = 0$.

We will conclude this section with the following corollary from Theorem 29.3, compare Definition 25.4.

Corollary 29.6. Let, under the assumptions of Theorem 29.3, $\nu = \mu^n$. Then $\mathrm{cond}(A,D,E) \leq n+cn^{16}$ for a constant c.

The bound $n+cn^{16}$ does not seem to be too large. Both Theorem 29.5 and Corollary 29.6 indicate that, for any algorithm for Det(n), chopping may cause a relatively large error due to the large range of the output values of the det X in the domain D. Example 29.1 demonstrates the growth of the relative output error for Det(n).

30. Summary of the Bounds on the Bit-Time of Computations in Linear Algebra; Acceleration of Solving a System of Linear Equations Where High Relative Precision of the Output is Required.

Comparing the upper and lower bounds on the bit-time in Theorems 24.2 and 24.3 for MM(n); Corollary 27.2 and Remark 27.1 for SLE(n); Corollary 28.2 and Remark 28.1 for MI(n); Corollary 29.4 and Theorem 29.5 for Det(n) and taking into account (18.3), we can see that the lower bounds are roughly the same in the cases of MM(n), SLE(n), and MI(n) (in the domains defined by the parameters $\mu(n)$, $\mu'(n)$, $\mu''(n)$, and $\nu(n)$ having the same order of magnitudes) and that they increase roughly n times in the case of Det(n). In each case the upper bound exceeds the lower bound roughly $n^{\omega-2+\varepsilon}$ times where ω is the exponent of MM over rational constants and ε is arbitrary positive number.

In this section we will show how to improve over the above upper bounds in the case of SLE(n) by modifying our previous computational schemes. In particular we will allow to compute with chopping to s bits at some steps and to s^* bits at other steps.

Theorem 30.1. Let E,μ,μ',ν be positive parameters such that $\mu\nu \geq 1$, $\mu'\nu > E$. Then there exists an algorithm A that solves a system of linear equations (21.1) in the domain D defined by (27.1),(27.2) with a prescribed precision E such that simultaneously $\mathrm{bs}(A,D,E) = O(n^2 \mathrm{maximum}\{s,s^*\})$ and $\mathrm{bt}(A,D,E) = O(n^{\omega+\varepsilon}\mathrm{bt}(s) + n^2 k\ (\mathrm{bt}(s^*)+\mathrm{bt}(s)))$ where ω,ε, and bt(s) are defined as in Theorem 23.6, provided that

$$\left.\begin{array}{l}\tilde{s} = s-1-\lceil \log(2cn^{15}\mu^2\nu^2)\rceil > 1, \\[2em] k > \log(\mu'/(\mu E))/(s-1-\log(2cn^{15}\mu^2\nu^2)),\ s^* \geq 2s+2\log(cn^{14}\mu\nu)\end{array}\right\} \quad (30.1)$$

for a positive constant c.

The latter bound is a substantial improvement over the bound $O(n^{\omega+\varepsilon}\ \mathrm{bt}(\log(\mu\mu'\nu^2/E)))$ of Corollary 27.2 in the cases where a high precision of the approximations to $\underline{x}$ is required, specifically where $\log(\mu\,\nu) = o(\log(\mu'\,\nu/E))$. Indeed, choose k and s^* barely satisfying the inequalities of (30.1). To keep the value $n^{\omega-2}bt(s)+k(bt(s^*) + bt(s))$ smaller, choose, say, $s = 1+2\log(2cn^{15}\mu^2\nu^2)$. Then

$$k(bt(s^*)+bt(s)) = O(\log(\mu'/(\mu E))bt(\log(n\,\mu\,\nu))/\log(n\,\mu\,\nu)),$$

$$bt(A,D,E) = O(n^{\omega+\varepsilon}bt(\log(\mu\,\nu)) + n^2\log(\mu'/(\mu\,E))bt(\log(n\,\mu\,\nu))/\log(n\,\mu\,\nu))).$$

Recall (18.3) and the assumption that $\log(\mu\,\nu) = o(\log(\mu'\,\nu/E))$ to see an improvement over the bound of Corollary 27.2. ∎

The bound of Theorem 30.1 is _near_ _sharp_ (within the factor $O(bt(\log(\mu\,\nu))/\log(\mu\,\nu)))$ in the cases where $\log(\mu'\,\nu/E) > cn^{\omega+\varepsilon-2}$ for positive constants c and ε, compare Remark 27.1. In any case, the mere existence of an improvement over the upper bound of Corollary 27.2 looks interesting.

Proving Theorem 30.1 we will refer to the following well-known concept of _significant_ _binary_ _digits_ (_bits_) of an approximation to a real number.

Definition _30.1_. Let a real number u and its approximation u^* be represented as the fixed point binary numbers

$$u = \sum_{k=-\infty}^{q} u_k 2^k, \quad u^* = \sum_{k=-\infty}^{q} u_k^* 2^k$$

where u_k and u_k^* are either 0 or 1 for each k. Then for an integer r (which is not necessarily positive) the approximation u^* is said to have _significant_ _bits_ to within 2^{r-1} (or, equivalently, the bits u_k^* for $k=r,r+1,\ldots,q$ are called _significant_) if $|u - u^*| \le 2^{r-1}$. (For instance, if $u_q=1$, $u_k=0$ for $q-4<k<q$, $u_q^* = 0$, $u_k^* = 1$ for $q-4<k<q$, then the bits $q,q-1,q-2$ are significant because in that case $|u - u^*| \le 2^{q-3}$.)

Proof _of_ _Theorem_ _30.1_. We will rely on the well-known method of iterative improvement of the solution to a system of linear equations, see [A], p. 469; [Wil], pp. 155-156. Let the desired algorithm A initially proceed as in Section 27. First apply Algorithm 20.2 with chopping to s bits throughout the computation, where s satisfies (30.1) and evaluate the approximations Q(s) and R(s) to the QR-factors of X. Then evaluate an approximation $\underline{z}(s)$ to the solution $\underline{z}$ to (21.1) by back substitution as in Section 21, see (27.5)-(27.7) (again with chopping to s bits). To explain the next step of the algorithm A, repeat the proof of Theorem 27.1 and derive that

$$||\underline{z} - \underline{z}(s)|| \le c\,h\,n^{15}\,\mu\,\nu^2\,||\underline{v}|| \tag{30.2}$$

where c is a constant such that (27.10) holds and $h=h(s)=2^{1-s}$, see Notation 23.1. This means that all entries of $\underline{z}(s)$ have significant digits to within 2^{-s_0} where

$$s_0 = s-1- \lceil \log(c\, n^{15}\, \mu\, \nu^2\, ||\underline{v}||) \rceil. \qquad (30.3)$$

Now chop all entries of $\underline{z}(s)$ preserving only the above significant digits. Let $\underline{z}^{(0)}$ designate the vector of the resulting approximations. (30.2) will still hold if we replace $\underline{z}(s)$ by $\underline{z}^{(0)}$.

At this point the algorithm A has reduced the original problem to solving the system of linear equations in $\underline{y}$,

$$X\,\underline{y} = \underline{v}^{(1)},$$

such that

$$\underline{z} = \underline{z}^{(0)} + \underline{y},\ \underline{v}^{(1)} = \underline{v} - X\,\underline{z}^{(0)} = X(\underline{z} - \underline{z}^{(0)}). \qquad (30.4)$$

($\underline{v}^{(1)}$ is the <u>residual</u> <u>vector</u> of the given system (21.1). Here we assume at first that the vector $\underline{v}^{(1)}$ has been computed with no error.)

As follows from (27.1),(30.2), and (30.4),

$$||\underline{v}^{(1)}|| \leq c\, h\, n^{15}\, \mu^2\, \nu^2\, ||\underline{v}||. \qquad (30.5)$$

The system $X\underline{y} = \underline{v}^{(1)}$ has the same matrix X as the original system $X\underline{z} = \underline{v}$ does, see (21.1); the algorithm A is recursively applied again and the same matrices Q(s) and R(s) are used again. The difference with the previous application of A is that, due to (30.5), the error bound of the computed solution to the system $X\underline{y} = \underline{v}^{(1)}$ is $c\, h\, n^{15}\, \mu^2\, \nu^2$ times less than the error bound (30.2). If $\underline{y}^{(0)}$ is the approximation to $\underline{y}$ computed by A (again with chopping to s bits), then the approximation $\underline{z}^{(1)} = \underline{z}^{(0)} + \underline{y}^{(0)}$ to $\underline{z}$ has $\tilde{s}$ additional significant digits where $\tilde{s}$ is defined by (30.1). The same improvement by adding new $\tilde{s}$ significant binary digits to the computed approximation to $\underline{z}$ takes place after each new iteration of the algorithm A performed with chopping to s bits, provided that at all steps the residual vectors are computed exactly. Then after k iteration steps of the algorithm A the error vector $\underline{\Delta}'(\underline{z},s,k)$ of the computed approximation to $\underline{z}$ will be bounded as follows,

$$||\, \underline{\Delta}'\ (\underline{z};s,k)|| \leq (c\, h\, n^{15}\, \mu^2\, \nu^2)^k\, ||\underline{v}||/\, \mu \leq (c\, h\, n^{15}\, \mu^2\, \nu^2)^k\, \mu'/\mu, \qquad (30.6)$$

compare (30.2),(30.5); that is, the error has been decreased $c\, h\, n^{15} \mu^2\, \nu^2$ times per one step. We should choose k and s such that

$$s > 1 + \log(cn^{15}\mu^2\nu^2);\ k > \log(\mu'/(\mu\, E))\ /\ (s - 1 - \log(c\, n^{15}\mu^2\, \nu^2\,)). \qquad (30.7)$$

The inequalities (30.7) follow if k and s satisfy (30.1). Actually even if we replace 2c by c in (30.1), the resulting bounds will still imply (30.7). On the other hand, (30.7) implies the upper bound E on the norm of the matrix of the output errors provided that all residuals have been computed with no error.

It remains to bound the errors of computing the residuals so that such errors would cause no substantial increase of the output errors. If the residual vector $\underline{v}^{(1)}$ is computed with the error vector $\underline{\Delta}^{(1)}$, then this contributes the error vector $X^{-1}\underline{\Delta}^{(1)}$ to the output approximation. We define k,s, and s^* satisfying (30.1) and require that the residual vector $\underline{v}^{(1)}$ be evaluated from (30.4) by the straightforward algorithm with chopping to s^* bits throughout that algorithm. The bound $\tilde{s} > 1$ in (30.1) implies that $||\underline{z}^{(0)}|| < 2||\underline{z}||$, so that we may use, say, $2||\underline{z}||$ at the place of $||\underline{z}^{(0)}||$ in our estimates. Similarly for the norms of the subsequent approximations to $\underline{z}$. Applying again Proposition 23.2, we derive the following bounds,

$$||\underline{\Delta}^{(1)}|| \leq 2h^* n^2 \mu\, ||\underline{z}|| + O(h^*)^2 \text{ where } h^* = 2^{1-s^*},$$

and consequently

$$||X^{-1}\underline{\Delta}^{(1)}|| \leq 2h^* n^2 \mu \nu\, ||\underline{z}|| + O(h^*)^2.$$

Following Assumption 26.3 we will ignore the higher order terms $O(h^*)^2$. Combining the latter inequality with (27.2) and (30.1), we derive that

$$||X^{-1}\underline{\Delta}^{(1)}|| \leq (c\ h\ n^{15}\mu^2\nu^2)^2\mu'/\mu.$$

Therefore the additional errors $X^{-1}\underline{\Delta}^{(1)}$ at most double our bound on the error norm of the approximation of $\underline{y}^{(0)}$ to $\underline{y}$, see (30.6) for k=2, which amounts to less than doubling the constant c in (30.6). Similarly we do at all of the k-1 subsequent iterations. Then we easily verify that the increase of the output errors due to the errors in computing the residuals with chopping to s^* bits amounts to less than doubling the factor c in (30.6) for all k, so that (30.1), indeed, assures the prescribed precision E of the output. Therefore the total bit-time of performing all of the k iterations and the preliminary step of the QR-factorization is bounded as we required in Theorem 30.1. ■

The estimates of Theorem 30.1 can be improved in the case where $s^* \geq 2s + \log n + 1$ and where all entries of the given matrix X and of the vector $\underline{v}$ are integers whose upper bounds are defined by (27.1) and (27.2). Since the successive approximations $\underline{z}^{(0)}, \underline{y}^{(0)}, \ldots$ are computed with chopping to s bits, see (30.1) and (30.3), we may evaluate the residuals at all steps exactly (that is, with infinite precision) if we compute them with $s' = 2s + \lceil \log n \rceil$ bits. Then we will improve the upper bounds on bt(A,D,E) in Theorem 30.1 by substituting $bt(s')$ for $bt(s^*)$.

Even if we relax one of the above requirements, that is, the requirement that the entries of the vector $\underline{v}$ be integers (but presere the requirement that the entries of the matrix X be integers), still we may assure the required precision computing with s bits throughout except that s' bits should be used while the residuals are computed and furthermore s^* bits ought to be used where $X\underline{z}^{(0)}$ is subtracted from $\underline{v}$, $X\underline{y}^{(0)}$ from $\underline{y}$, and so on. In that case we need only $O(n\ bt(s^*))$ additional bit-operations comparing with the previous case, where all input entries were integers.

Finally, if the matrix X is symmetric and sparse, so that its product by a vector is computed in O(n) arithmetical operations, then it is effective to use the unitary transformation of X into a tridiagonal matrix rather than into a triangular matrix R. Then the new approximations $\underline{z}^{(0)}$, $\underline{y}^{(0)}$,... will be computed from tridiagonal systems of linear equations involving O(n) arithmetical operations per system. This would enable us to replace the factors n^2 by n in the upper bounds on bt(A,D,E) in Theorem 30.1 (in the general case and in the special case of integer entries of the inputs). The extension of the latter approach to the case of nonsymmetric sparse matrices X is possible via the standard symmetrization of the system (21.1) by the premultiplication by X^H, which can be kept implicit at the stage of computing the residual vectors.

PART 3. THE SPEED-UP OF THE MULTIPLICATION OF MATRICES OF A FIXED SIZE.

Summary. The algorithms for NXN MM in $O(N^{2.496})$ arithmetics presented in Part 1 become superior over the straightforward algorithm only for enormously large N, compare Remark 6.4 and the proof of Theorem 7.2. Those asymptotically fastest algorithms become extremely inefficient if they are applied to the multiplication of matrices of moderate sizes. Thus we need to devise some special algorithms for the latter problem. Their design and analysis with some natural extensions to the study of other related computational problems will be the subject of this part of the monograph.

In particular we will recall the bilinear algorithms and λ-algorithms of the currently least ranks and λ-ranks for nXn MM for all specific moderate n (see Sections 31 and 33), will consider the extensions to the classes of commutative quadratic algorithms and λ-algorithms (in Sections 32 and 33) and of the general arithmetical algorithms and λ-algorithms (in Section 34), and will recall some correlation between these algorithms and bilinear ones. We will show that the commutative quadratic λ-algorithms can be surprinsingly efficient already for multiplication of matrices of small sizes (and also for the evaluation of polynomials, see Example 34.2 in Section 34) although such λ-algorithms (and even more general arithmetical algorithms and λ-algorithms, see Section 34) help little for the asymptotic acceleration of MM. We will recall some lower bounds in Sections 35-38 for the sake of completeness and in order to demonstrate some typical algebraic techniques for that problem (basic active substitution argument, (linear) independence argument, reduction modulo the ideals defined by cubic terms, matrix-vector representation and tensor representation of quadratic algorithms, matrix representation of linear algorithms, duality and some other techniques of reduction of problems and algorithms to each other). Finally we will introduce a quantity for measuring asynchronicity of linear and bilinear algorithms in Section 39, and will demonstrate a new promising extension of the class of arithmetical λ-algorithms in Section 40.

31. The Currently Best Upper Bounds on the Rank of the Problem of MM of Moderate Sizes.

We will start this section with the table of the currently best upper bounds on the ranks $\rho(n,n,n)$ of the problems of nXn MM for some small n (over all rings of constants).

Table 31.1

n	upper bound on the rank $\rho(n,n,n)$	Paper
2	7	[St69]
3	21	[La]
4	49	[St69]
5	103	[Scha]

The bounds in the cases n=6,8,9,12,16,18 can be obtained from the recursive algorithms derived from Strassen's algorithm for n=2 and Laderman's algorithm for n=3. Similarly for several other sizes n with the additional option of the reduction to the multiplication of square and rectangular block-submatrices of appropriate sizes, compare [Pr76]. However, for n=10, n=14, and for all n ranging from 20 to 10^{20}, the current least upper bounds on $\rho(n,n,n)$ rely on the next theorem and on its combination with Proposition 2.2.

Theorem 31.1. For all even N and for all rings of constants F that contain the inverse of N/2, the following bounds hold,

$$\rho_F(N,N,N) \leq N^3/3 + (15/4)N^2 + (97/6)N + 20, \tag{31.1}$$

$$\rho_F(N,N,N) \leq N^3/2 + 9N^2/4 - (3/2)N, \tag{31.2}$$

$$\rho_F(N,N,N) \leq N^3/3 + 4N^2 + (35/3)N, \tag{31.3}$$

$$\rho_R(N,N,N) \leq N^3/3 + 4n^2 + (26/3)N. \tag{31.4}$$

Here R is the field of real numbers.

Remark 31.1. (31.2) has been taken from [P80]; it defines the current least upper bounds on $\rho_F(n,n,n)$ for all rings F for n=10 and n=14, that is, 710 and 1792, respectively. (31.3) and (31.4) have been taken from [P81]; (31.4) defines the current least upper bounds on $\rho_R(N,N,N)$ for N ranging from 20 to 32, and was actually proven in [P81] over the ring of constants that is larger than R. For N=34 and for greater N, (31.1) gives superior bounds.

In the next table we represent the currently best upper bounds M(n) on $\rho_F(n,n,n)$ and the associated exponents for some moderate n.

Table 31.2

n	M(n)	log M(n)/log n
20	4440	2.8035
24	7120	2.7912
30	12860	2.7819
36	21014	2.7775
40	28000	2.7759
42	32010	2.7755
44	36386	2.77522
46	41144	2.775093
48	46300	2.775082
50	51870	2.77516
52	57870	2.77532

Next we will prove (31.1) by constructing bilinear algorithms of the required rank for NXN MM for all even N. We will define those algorithms in the form of trilinear decompositions using the techniques of trilinear aggregating (TA) along the line of Section 13. Now the algorithmic design becomes more tricky, however, because we would not be satisfied with obtaining λ-algorithms and thus will not use the auxiliary parameter λ. (Actually the design is still simpler than ones of [P78],[P80], and [P81] due to the benefits of Propositions 31.2 and 31.3 of this section, compare Remark 13.1.)

At first we will demonstrate the main techniques in a similar construction of algorithms for a problem of Disjoint MM. We will rely on the following table (compare Table 13.1)

Aggregating Table 31.3

a_{ij}	b_{jk}	c_{ki}
u_{jk}	v_{ki}	w_{ij}
x_{ki}	y_{ij}	z_{jk}

Table 31.3 defines TA for the problem represented by Tr(ABC)+Tr(UVW)+Tr(XYZ).

We will transform the given problem of Disjoint MM similarly to our transformations in Section 13. Namely we will write

$$\sum_r d_{rs} = \sum_s d_{rs} = 0 \tag{31.5}$$

and will require that (31.5) hold for d_{rs} standing for each of $a_{ij}, b_{jk}, c_{ki}, u_{jk}, v_{ki}, w_{ij}, x_{ki}, y_{ij}, z_{jk}$.

We will use the next two auxiliary results where $\sum_r$ and $\sum_r^*$ designate the sums in r ranging from 0 to N-1 and from 1 to N-1, respectively.

Proposition 31.2. Let

$$\sum_r g_r = 0, \sum_r h_r = 0. \tag{31.6}$$

Then

$$\sum_r g_r h_r = \sum_r^* g_r^* h_r \tag{31.7}$$

where

$$g_s^* = g_s + \sum_r^* g_r, \ s=1,2,\ldots,N-1. \tag{31.8}$$

Proof. $\sum_r g_r h_r = \sum_r^* g_r h_r + g_0 h_0$. Substitute $h_0 = -\sum_r^* h_r$ (see (31.6)); obtain that $\sum_r g_r h_r = \sum_r^* (g_r - g_0) h_r$. Substitute $g_0 = -\sum_r^* g_r$ (see (31.6)), apply (31.8), and derive (31.7). ∎

Proposition 31.3. The system of linear equations (31.8) can be solved in $g_1, g_2, \ldots, g_{N-1}$ over any ring F that contains $g_1^*, g_2^*, \ldots, g_{N-1}^*$, and the inverse of N among its elements.

Proof. Subtract the equation

$$g_{N-1}^* = g_{N-1} + \sum_r^* g_r \tag{31.9}$$

from all other equations $g_s^* = g_s + \sum_r^* g_r$ of (31.8) and derive that $g_s = g_{N-1} + g_s^* - g_{N-1}^*$, s=1,2,...,N-2. It remains to find g_{N-1}. Substitute the latter expressions for $g_1, g_2, \ldots, g_{N-2}$ into (31.9) and derive that

$$Ng_{N-1} = Ng_{N-1}^* - \sum_r^* g_r^*, \ g_{N-1} = g_{N-1}^* - \sum_r^* g_r^*/N.$$

Therefore the system $g_s^* = g_s + \sum_r^* g_s$ can be solved over F in g_s, s=1,...,N-1. ∎

Remark 31.2. Similarly it is possible to deduce from (31.6) that

$$\sum_r g_r h_r = \sum_r^* g_r h_r^* \text{ where } h_s^* = h_s + \sum_r^* h_r, \ s=1,\ldots,N-1,$$

and that the latter system can be solved in $h_1, h_2, \ldots, h_{N-1}$ over any ring that contains $h_1^*, \ldots, h_{N-1}^*$, and N^{-1} among its elements.

Successively applying Propositions 31.2 and 31.3 and using Remark 31.2 we derive that the linear equations (31.5) transform the original problem of Disjoint MM $(m,n,p) \oplus (n,p,m) \oplus (p,m,n)$ into the problem $(m-1,n-1,p-1) \oplus (n-1,p-1,m-1) \oplus (p-1,m-1,n-1)$. Table 31.3 and the relation (31.5) define the trilinear decomposition where all correction terms have been canceled except for the terms $a_{ij}v_{ki}z_{jk}$, $x_{ki}b_{jk}w_{ij}$, $u_{jk}y_{ij}c_{ki}$ for all i,j,k, which we will call <u>diagonal correction terms</u>, and for the terms $a_{ij}y_{ij}w_{ij}$, $u_{jk}b_{jk}z_{jk}$, $x_{ki}v_{ki}c_{ki}$ for all i,j,k, which we will call <u>antidiagonal correction terms</u>. To get rid of the 3mnp diagonal correction terms, use special canceling procedure due to [P78]. Namely, sum the eight aggregates defined by Table 31.3 and by the next seven aggregating tables where $\tilde{i} = i+m$, $\tilde{j} = j+n$, $\tilde{k} = k+p$. (Compare Tables 13.1 and 13.2.)

Aggregating Table 31.3

a_{ij}	b_{jk}	c_{ki}
u_{jk}	v_{ki}	w_{ij}
x_{ki}	y_{ij}	z_{jk}

Aggregating Table 31.4

$-a_{ij}$	$b_{j\tilde{k}}$	$-c_{\tilde{k}i}$
$u_{\tilde{j}k}$	v_{ki}	$w_{i\tilde{j}}$
$x_{k\tilde{i}}$	$y_{\tilde{i}j}$	z_{jk}

Aggregating Table 31.5

$a_{i\tilde{j}}$	$b_{\tilde{j}k}$	c_{ki}
$-u_{jk}$	$v_{k\tilde{i}}$	$-w_{\tilde{i}j}$
$x_{\tilde{k}i}$	y_{ij}	$z_{j\tilde{k}}$

Aggregating Table 31.6

$a_{\tilde{i}j}$	b_{jk}	$c_{k\tilde{i}}$
$u_{j\tilde{k}}$	$v_{\tilde{k}i}$	w_{ij}
$-x_{ki}$	$y_{i\tilde{j}}$	$-z_{\tilde{j}k}$

Aggregating Table 31.7

$a_{\tilde{i}\tilde{j}}$	$b_{\tilde{j}\tilde{k}}$	$c_{\tilde{k}\tilde{i}}$
$u_{\tilde{j}\tilde{k}}$	$v_{\tilde{k}\tilde{i}}$	$w_{\tilde{i}\tilde{j}}$
$x_{\tilde{k}\tilde{i}}$	$y_{\tilde{i}\tilde{j}}$	$z_{\tilde{j}\tilde{k}}$

Aggregating Table 31.8

$-a_{\tilde{i}\tilde{j}}$	$b_{\tilde{j}k}$	$-c_{k\tilde{i}}$
$u_{j\tilde{k}}$	$v_{\tilde{k}\tilde{i}}$	$w_{\tilde{i}j}$
$x_{\tilde{k}i}$	$y_{i\tilde{j}}$	$z_{\tilde{j}\tilde{k}}$

Aggregating Table 31.9

$a_{\tilde{i}j}$	$b_{j\tilde{k}}$	$c_{\tilde{k}\tilde{i}}$
$-u_{\tilde{j}\tilde{k}}$	$v_{\tilde{k}i}$	$-w_{i\tilde{j}}$
$x_{k\tilde{i}}$	$y_{\tilde{i}\tilde{j}}$	$z_{\tilde{j}k}$

Aggregating Table 31.10

$a_{i\tilde{j}}$	$b_{\tilde{j}\tilde{k}}$	$c_{\tilde{k}i}$
$u_{\tilde{j}k}$	$v_{k\tilde{i}}$	$w_{\tilde{i}\tilde{j}}$
$-x_{\tilde{k}\tilde{i}}$	$y_{\tilde{i}j}$	$-z_{j\tilde{k}}$

Note that each diagonal correction term now appears in Tables 31.3-31.10 exactly twice and with the opposite signs (+ and -) so that all such terms cancel themselves in the sum of the eight aggregates of those eight tables.

The sum of all principal terms of Tables 31.3-31.10 for all i,j,k is identically the trilinear form

$$T = \sum_{i=0}^{2m-1} \sum_{j=0}^{2n-1} \sum_{k=0}^{2p-1} (a_{ij}b_{jk}c_{ki} + u_{jk}v_{ki}w_{ij} + x_{ki}y_{ij}z_{jk}),$$

which represents the problem of Disjoint MM, $(2m,2n,2p) \oplus (2n,2p,2m) \oplus (2p,2m,2n)$. Such a problem shrinks when we impose the equation (31.5). Furthermore we will extend the equation (31.5) and will require that

$$\sum_r d_{rs} = \sum_r d_{\tilde{r}s} = \sum_r d_{r\tilde{s}} = \sum_r d_{\tilde{r}\tilde{s}} = \sum_s d_{rs} = \sum_s d_{\tilde{r}s} = \sum_s d_{r\tilde{s}} = \sum_s d_{\tilde{r}\tilde{s}} = 0 \qquad (31.10)$$

where each r and each s stand for i,j, or k, and d stands for a,b,c,u,v,w,x,y, or z.

This reduces the above problem of Disjoint MM to the following problem of a smaller size, $(2m-2,2n-2,2p-2) \circledast (2n-2,2p-2,2m-2) \circledast (2p-2,2m-2,2n-2)$, see Propositions 31.2 and 31.3 and Remark 31.2, but, on the other hand, this cancels all of the remaining correction terms except for 8(mn+np+pm) antidiagonal ones, that is, except for the mn+np+pm terms $pa_{ij}y_{ij}w_{ij}, mu_{jk}b_{jk}z_{jk}, nx_{ki}v_{ki}c_{ki}$ (for all i,j,k) from Table 31.3 and for the 7(mn+np+pm) similar terms from other seven tables. Therefore $\rho(T) \leq 8(mnp+mn+np+pm)$, counting the 8mnp aggregates (one per each triplet (i,j,k) in each of the eight tables) and the 8(mn+np+pm) antidiagonal correction terms and assuming that (31.10) holds.

Next examine the following sum of 8 antidiagonal correction terms.

$$p(a_{ij}y_{ij}w_{ij} - a_{ij}y_{\tilde{i}j}w_{i\tilde{j}} - a_{i\tilde{j}}y_{ij}w_{\tilde{i}j} + a_{\tilde{i}j}y_{i\tilde{j}}w_{ij} +$$

$$a_{\tilde{i}\tilde{j}}y_{\tilde{i}\tilde{j}}w_{\tilde{i}\tilde{j}} - a_{\tilde{i}\tilde{j}}y_{i\tilde{j}}w_{\tilde{i}j} - a_{\tilde{i}j}y_{\tilde{i}\tilde{j}}w_{i\tilde{j}} + a_{i\tilde{j}}y_{\tilde{i}j}w_{\tilde{i}\tilde{j}}).$$

Note that such a sum represents 2X2 MM for each fixed pair (i,j). (To verify that property, designate $X_{00} = pa_{ij}$, $X_{10} = pa_{i\tilde{j}}$, $X_{01} = pa_{\tilde{i}j}$, $X_{11} = pa_{i\tilde{j}}$, $Y_{00} = y_{ij}$, $Y_{10} = y_{i\tilde{j}}$, $Y_{01} = y_{\tilde{i}j}$, $Y_{11} = y_{\tilde{i}\tilde{j}}$, $Z_{00} = w_{ij}$, $Z_{10} = -w_{i\tilde{j}}$, $Z_{01} = -w_{\tilde{i}j}$, $Z_{11} = w_{\tilde{i}\tilde{j}}$.) Apply Strassen's algorithm and reduce the number of terms from 8 to 7. Do this for every pair (i,j) and similarly for every pair (j,k) and for every pair (k,i) and reduce the total number of remaining correction terms to 7(mn+np+pm).

Summarizing obtain the following bilinear algorithm (mapping) for Disjoint MM (for arbitrary m,n,p).

$$(2m-2,2n-2,2p-2) \circledast (2n-2,2p-2,2m-2) \circledast (2p-2,2m-2,2n-2) \leftarrow (8mnp+7(mn+np+pm)) \odot (1,1,1).$$

If m=n=p, this implies

$$3 \odot (2n-2,2n-2,2n-2) \leftarrow (8n^3+21n^2) \odot (1,1,1) \text{ for all } n. \qquad (31.11)$$

To extend the latter algorithm (mapping) to an algorithm for MM, compress the three disjoint problems of MM into one problem of MM by identifying the input variables within each of the three columns of the aggregating tables as follows,

$$a_{rs} = u_{rs} = x_{rs},\ b_{rs} = v_{rs} = y_{rs},\ c_{rs} = w_{rs} = z_{rs} \text{ for all r and s.} \qquad (31.12)$$

Under these restrictions, Tables 31.3-31.10 turn into the eight compressed aggregating tables as follows (we will explicitly present only two compressed tables).

Compressed Table 31.3

a_{ij}	b_{jk}	c_{ki}
a_{jk}	b_{ki}	c_{ij}
a_{ki}	b_{ij}	c_{jk}

Compressed Table 31.4

$-a_{ij}$	$b_{j\tilde{k}}$	$-c_{\tilde{k}i}$
$a_{\tilde{j}k}$	b_{ki}	$c_{i\tilde{j}}$
$a_{k\tilde{i}}$	$b_{\tilde{i}j}$	c_{jk}

We will define TA by the 8-tuples of Compressed Aggregating Tables 31.3-31.10 for all triplets i,j,k such that either $0 \le i \le j < k \le n-1$ or $0 \le k < j \le i \le n-1$. There are exactly $(n^3-n)/3$ such triplets. The sum of all principal terms of those $8(n^3-n)/3$ tables equals $Tr(ABC)-T^*$ where

$$Tr(ABC) = \sum_{i,j,k=0}^{2n-1} a_{ij}b_{jk}c_{ki}, \quad T^* = \sum_{i=0}^{n-1} (a_{ii}b_{ii}c_{ii} + a_{ii}b_{i\tilde{i}}c_{\tilde{i}i} +$$

$$a_{\tilde{i}i}b_{ii}c_{i\tilde{i}} + a_{i\tilde{i}}b_{\tilde{i}i}c_{ii} + a_{\tilde{i}\tilde{i}}b_{\tilde{i}\tilde{i}}c_{\tilde{i}\tilde{i}} + a_{\tilde{i}\tilde{i}}b_{\tilde{i}i}c_{i\tilde{i}} +$$

$$a_{i\tilde{i}}b_{\tilde{i}\tilde{i}}c_{\tilde{i}i} + a_{\tilde{i}i}b_{i\tilde{i}}c_{\tilde{i}\tilde{i}}).$$

Under the equations (31.10),(31.12), $Tr(ABC)$ represents the problem of $(2n-2) \times (2n-2)$ MM, see Propositions 31.2 and 31.3 and Remark 31.2.

On the other hand, the sum of all aggregates of the Compressed Aggregating Tables 31.3-31.10 equals $Tr(ABC) - T^* - T_0 - T_1$. Here T_0 is the sum of all $8n^2$ antidiagonal correction terms in Compressed Aggregating Tables 31.3-31.10 for all i,j,k and T_1 is the sum of all remaining correction terms. (We will see that the equations (31.10) play a different role now due to the new range of i,j,k.) Let T_0' designate the sum of all $8n$ antidiagonal correction terms in T_0 that are in the form $ra_{ii^*}b_{jj^*}c_{kk^*}$ where i,i^*,j,j^*,k,k^* coincide with each other modulo n and where r is an integer.

Let us show that T_1 can be represented as the sum of 20n terms of such a special form. First consider Compressed Table 31.3 for the triplets (i,i,k) for all $k \ne i$. Only the sum of the nondiagonal and of the nonantidiagonal correction terms of that table contributes to T_1. That sum is equal to

$$\sum_i \sum_{k \ne i} (a_{ii}b_{ii}(c_{ki} + c_{ik}) + a_{ii}(b_{ki} + b_{ik})c_{ii} + (a_{ki} + a_{ik})b_{ii}c_{ii}),$$

not counting the terms containing at least 2 of the factors a_{ik}, b_{ik}, c_{ik}, a_{ki}, b_{ki}, c_{ki}. (The latter terms are canceled when they are summed with other correction terms and when the equations (31.10) are applied.) Since the above sum is not in all k but only in $k \neq i$, the equations (31.10) do not cancel that sum but turn it into $-6 \sum_i a_{ii}b_{ii}c_{ii}$. Similarly Compressed Table 31.7 contributes (to T_1) the correction terms whose sum equals $-6 \sum_i a_{\tilde{i}\tilde{i}}b_{\tilde{i}\tilde{i}}c_{\tilde{i}\tilde{i}}$. Therefore the rank of the sum of all correction terms contributed to T_1 by those two tables is only 2n. Furthermore we may also add here the terms $a_{ii}b_{ii}c_{ii}$ and $a_{\tilde{i}\tilde{i}}b_{\tilde{i}\tilde{i}}c_{\tilde{i}\tilde{i}}$ from T^* and the terms $(n-1)(a_{ii}b_{ii}c_{ii} + a_{\tilde{i}\tilde{i}}b_{\tilde{i}\tilde{i}}c_{\tilde{i}\tilde{i}})$ from T_0' and still have the rank of the resulting trilinear form bounded from above by 2n.

Similarly Compressed Aggregating Tables 31.4-31.6 contribute the following correction terms to T_1,

$$a_{ii}b_{\tilde{i}i}(-c_{\tilde{i}i} + c_{ii}),\ -(a_{\tilde{i}i} + a_{i\tilde{i}})\, b_{\tilde{i}i}c_{i\tilde{i}},\ a_{ii}(b_{i\tilde{i}} + b_{ii})\, c_{i\tilde{i}};$$

$$-a_{i\tilde{i}}b_{ii}(c_{ii} + c_{i\tilde{i}}),\ (-a_{ii} + a_{\tilde{i}i})\, b_{ii}c_{\tilde{i}i},\ a_{i\tilde{i}}(b_{\tilde{i}i} + b_{i\tilde{i}})c_{\tilde{i}i};$$

$$-a_{\tilde{i}i}\, b_{i\tilde{i}}(c_{i\tilde{i}} - c_{\tilde{i}i}),\ -(a_{i\tilde{i}} - a_{ii})\, b_{i\tilde{i}}c_{ii},\ -a_{\tilde{i}i}(b_{ii} + b_{\tilde{i}i})c_{ii}.$$

The sum of those terms for a fixed i has rank 9. Note that the rank of that sum will not increase if we add the antidiagonal terms $(n-1)a_{ii}b_{\tilde{i}i}c_{\tilde{i}i}$, $(1-n)a_{i\tilde{i}}b_{ii}c_{\tilde{i}i}$, $(n-1)a_{\tilde{i}i}\,b_{i\tilde{i}}c_{i\tilde{i}}$ from T_0' and also the terms $a_{\tilde{i}\tilde{i}}b_{\tilde{i}i}\,c_{i\tilde{i}}$, $a_{i\tilde{i}}b_{\tilde{i}\tilde{i}}c_{\tilde{i}i}$, $a_{\tilde{i}i}\,b_{i\tilde{i}}c_{\tilde{i}\tilde{i}}$ from T^*. Summing in all i we obtain a trilinear form of rank at most 9n.

Similarly sum all correction terms of T_1 generated by Compressed Aggregating Tables 31.8-31.10 with all antidiagonal correction terms from T_0 that are of the form $(n-1)a_{\tilde{i}\tilde{i}}b_{\tilde{i}\tilde{i}}c_{i\tilde{i}}$, $(1-n)a_{\tilde{i}i}\,b_{\tilde{i}\tilde{i}}c_{i\tilde{i}}$, $(n-1)a_{i\tilde{i}}b_{\tilde{i}i}\,c_{\tilde{i}i}$ and with all terms $a_{ii}b_{i\tilde{i}}c_{\tilde{i}i}$, $a_{\tilde{i}i}\,b_{ii}c_{i\tilde{i}}$, $a_{i\tilde{i}}b_{\tilde{i}i}\,c_{ii}$ of T^* and note that the rank of the resulting trilinear form is at most 9n. Therefore the rank of $T^* + T_1 + T_0'$ is at most 20n.

We decompose $T_0 - T_0'$ as the sum of n^2-n trilinear forms each representing 2X2 MM as we did devising the algorithm (31.11). Applying Strassen's algorithm, we reduce that sum to $7(n^2-n)$ terms.

Summarizing we have $(8/3)(n^3-n)$ aggregates, $7(n^2-n)$ terms in the decomposition of $T_0 - T_0'$, and 20n terms in the decomposition of $T^* + T_1 + T_0'$, that is, a total of

$$(8/3)(n^3-n) + 7(n^2-n) + 20n = (8/3)n^3 + 7n^2 + (31/3)n$$

terms in the trilinear form that represents the problem $(2n-2,2n-2,2n-2)$. Substitute $N=2n-2$, $2n=N+2$ and obtain that

$$\rho(N,N,N) \leq (1/3)(N+2)^3 + (7/4)(N+2)^2 + (31/6)(N+2) =$$

$$(1/3)N^3 + (15/4)N^2 + (97/6)N + 20. \qquad \blacksquare$$

We will conclude this section with presenting (without the proofs) the following three results, which show that the straightforward reduction of rectangular MM to square MM and to matrix-vector multiplication is not always the most efficient way to solve the problem.

Theorem 31.4 [HK]. $\rho(n,2,n) \leq n(3n+1)/2$ for all n over all rings of constants.

Theorem 31.5 [BD76]. $\rho_F(n,r(n),n) \leq n^2+o(n)$ as $n \to \infty$ provided that $\lim_{n \to \infty} \log r(n)/\log n = 0$ and that 2 has the inverse in F.

Theorem 31.6 [C]. Let $n \to \infty$. Then over all rings of constants

$$\rho(n,n,n^{\alpha}) = O(n^2(\log n)^2) \text{ for } \alpha = 2 \log 2/(5 \log 5) = 0.1722\ldots \text{ and}$$

$$\rho(n,n,n^{\beta}) = O(n^{2+\quad}) \text{ for } \beta = 0.197 \text{ and for all positive } \quad .$$

32. Commutative Quadratic Algorithms for MM.

Our efforts to reduce the ranks of bilinear algorithms for MM have been justified by Theorem 2.1 in the cases where we sought asymptotically fast methods for MM. If we consider a problem of MM of a fixed moderate size, then we naturally come back to minimizing the arithmetical complexity and/or the bit-complexity of more general algorithms for that problem.

The next simple example demonstrates the usefulness of the extension of the class of bilinear algorithms. Suppose we need evaluate x^2-y^2. We may use the straightforward algorithm but we may substitute 1 addition for 1 multiplication computing by the formula $x^2-y^2 = (x+y)(x-y)$ provided that $xy=yx$. The latter commutativity condition does not hold if x and y are matrices so that the above algorithm cannot be used in that case.

In this section we will assume that the multiplication of the input variables is commutative, extend the class of bilinear algorithms for MM to the class of commutative quadratic algorithms (see Definition 32.1 below), and will analyse the resulting acceleration of MM. Note that the acceleration will not be asymptotic because we cannot substitute matrices for variables and extend Theorem 2.1 if we require the commutativity of multiplication.

We will start with an example of a commutative quadratic algorithm for $m \times 2 \times p$ MM for moderate m and p that involves only $mp+m+p$ multiplications and $4mp$ additions and subtractions.

Example 32.1.

$$x_{i0}y_{0k} + y_{i1}y_{1k} = (x_{i0} + y_{1k})(x_{i1} + y_{0k}) - q_i(X) - s_k(Y) \tag{32.1}$$

where

$$q_i(X) = x_{i0}x_{i1},\ s_k(Y) = y_{1k}y_{0k} \tag{32.2}$$

and i ranges from 0 to m-1; k ranges form 0 to p-1.

It is assumed that the multiplication of the indeterminates x and y is commutative, so that

$$y_{1k}x_{i1} = x_{i1}y_{1k} \text{ for all i and k.}$$

The latter equations do not hold if arbitrary matrices substitute for the indeterminates, so that the algorithm cannot be used recursively. On the other hand, as we could see, commutative algorithms can be useful for mXnXp MM for specific m,n, and p. Let us study that direction in some detail.

Definition 32.1, compare Definition 2.2. A commutative quadratic algorithm A for the evaluation of a set of quadratic forms

$$q_\gamma(V) = \sum_{\sigma,\sigma'} g(\gamma,\sigma,\sigma')\, v_\sigma v_{\sigma'},\quad \gamma = 0,1,\ldots,\Gamma-1,$$

is defined by the following set of quadratic identities in the v-variables,

$$q_\gamma(V) = \sum_{g=0}^{M-1} f''(\gamma,g)L_g(V)L_g'(V), \tag{32.3}$$

$$L_g(V) = \sum_\sigma f(\sigma,g)v_\sigma,\ L_g'(V) = \sum_{\sigma'} f'(\sigma',g)v_{\sigma'} \text{ for all g.} \tag{32.4}$$

The algorithm A successively evaluates $L_g(V), L_g'(V), L_g(V)L_g'(V)$ for all g and $q_\gamma(V)$ for all γ involving $\rho(A) = M$ nonlinear multiplications of $L_g(V)$ by $L_g'(V)$ for all g. (They are also called nonscalar multiplications.) $\rho(A)$ is called the rank of A. as(A) and sm(A) designate the number of additions and subtractions and of linear (scalar) multipications involved in A, respectively. The minimum rank in all commutative quadratic algorithms for the problem of the evaluation of a given set $\{q_\gamma(V)\}$ is called the commutative rank of that problem.

Remark 32.1. It is assumed that the algorithms and ranks are defined over a given ring of constants.

<u>Remark 32.2</u>. Every bilinear algorithm is a commutative quadratic algorithm for the evaluation of a set of bilinear forms, compare Definitions 2.2 and 32.1.

(32.1),(32.2) define a commutative quadratic algorithm A for (m,2,p), which is not bilinear. For that algorithm A,

$$\rho(A) = mp+m+p, \; as(A) = 4mp, \; sm(A) = 0. \tag{32.5}$$

The algorithm (32.1),(32.2) can be extended to MM problems of different sizes using the following simple facts.

<u>Proposition 32.1</u>. For every commutative quadratic algorithm A for mXnXp MM, it is possible to define commutative quadratic algorithms A^* for pXnXm MM; A' for mX(rn)Xp MM, A'' for (rm)XnXp MM; A''' for mXnX(rp) MM such that

$$\rho(A^*) = \rho(A), \; as(A^*) = as(A), \; sm(A^*) = sm(A), \tag{32.6}$$

$$\rho(A') = \rho(A'') = \rho(A''') = \rho(A)r, \; sm(A') = sm(A'') = sm(A''') = sm(A)r, \tag{32.7}$$

$$as(A') = as(A)r + mp, \; as(A'') = as(A''') = as(A)r. \tag{32.8}$$

<u>Proof</u>. Rewrite $\sum_j x_{ij}y_{jk}$ as $\sum_j x^*_{kj}y^*_{ji}$ where $x^*_{kj} = y_{jk}$ and $y^*_{ji} = x_{ij}$ and reduce the problems (m,n,p) to (p,n,m) and the algorithms A to A^* such that (32.6) holds. To devise the desired algorithm A', simply reduce the problem (rm,n,p) to r problems (m,n,p) and apply the algorithm A to each of them. Similarly derive the algorithms A'' an A'''. ∎

<u>Remark 32.3</u>. (32.1) extends the duality equation $\rho(m,n,p) = \rho(p,n,m)$, see Theorem 2.3, to the case of commutative quadratic algorithms. The duality equations $\rho(m,n,p) = \rho(n,p,m) = \rho(p,m,n)$ are not extendable to that case, however.

Applying Proposition 32.1 to the algorithm A of Example 32.1, compare (32.5)-(32.8), we obtain the following result.

<u>Corollary 32.2</u>. For all m,p, and r, there exists a commutative quadratic algorithm A' for the problem mX(2r)Xp MM such that

$$\rho(A') = (mp+m+p)r, \; as(A') = mp(4r+1), \; sm(A') = 0.$$

Let us demonstrate how the number of additions and subtractions can be reduced even further if those operations in A' are rearranged in an appropriate order.

<u>Example 32.2</u>. A commutative quadratic algorithm for mX(2r)Xp MM is defined by the following identities,

$$\sum_{j=0}^{2r-1} x_{ij}y_{jk} = \sum_{h=0}^{r-1} (x_{i,2h} + y_{2h+1,k})(x_{i,2h+1} + y_{2h,k}) - q_i(X) - s_k(Y) \tag{32.9}$$

where

$$q_i(X) = \sum_{h=0}^{r-1} x_{i,2h}x_{i,2h+1}, \quad s_k(Y) = \sum_{h=0}^{r-1} y_{2h+1,k}y_{2h,k}, \qquad (32.10)$$

$i=0,1,\dots,m-1$; $k=0,1,\dots,p-1$. The equations (32.9) and (32.10) define a commutative quadratic algorithm A for $m\times(2r)\times p$ MM such that

$$\rho(A) = (mp+m+p)r, \quad as(A) = mp(3r+1) + (m+p)(r-1), \quad sm(A) = 0. \qquad (32.11)$$

In particular for $m=n=2r$ we have that

$$\rho(A) = 0.5\,n^3 + n^2, \quad as(A) = 1.5\,n^3 + 2n^2 - 2n, \quad sm(A) = 0. \qquad (32.12)$$

Remark 32.4. Example 32.2 was first published in [W68]. This example is a simple and natural variation of the previously developed schemes for the evaluation of polynomials with preconditioning, see [Mo],[P66],[K]. The present author gave a lecture about that algorithm at the seminar led by A.S. Kronrod, E.M. Landis, and G.M. Alel'son-Vel'skii as early as in January 1966.

As a matter of fact, the number of nonscalar multiplications in the algorithms of Examples 32.1 and 32.2 can be slightly reduced at the expense of a more substantial increase of the number of linear operations.

Example 32.3 [Wa] ($m\times2$ by $2\times m$ MM). Let $b_{ik}(X,Y)$ be as in Example 32.1. Let i and k range from 0 to m-1 and h range from 1 to m-1. Let

$$r_{ik}(X,Y) = (x_{i0} + y_{1k})(y_{0k} + x_{i1}),$$

$$p_i(X,Y) = (x_{i0} - y_{1i})(y_{0i} - x_{i1}),$$

$$q_h(X,Y) = (x_{00} - y_{1h})(y_{0h} - x_{01}),$$

Then

$$s_i^*(X,Y) = 0.5(r_{ii}(X,Y) - p_i(X,Y)) = y_{1i}y_{0i} + x_{i0}x_{i1},$$

$$t_h(X,Y) = 0.5(r_{0h}(X,Y) - q_h(X,Y)) = y_{1h}y_{0h} + x_{00}x_{01},$$

$$t_0(X,Y) = 0.5(r_{00}(X,Y) - p_0(X,Y)) = x_{00}x_{01} + y_{10}y_{00},$$

$$s_i(X) = s_i^*(X,Y) - t_i(X,Y) = x_{i0}x_{i1} - x_{00}x_{01},$$

$$x_{i0}y_{0k} + x_{i1}y_{1k} = r_{ik}(X,Y) - s_i(X) - t_k(X,Y).$$

The latter algorithm involves m^2+2m-1 nonscalar multiplications, 2m scalar multiplications by 1/2 and $4m^2+7m-2$ additions and subtractions. The reader may easily extend the algorithm to mXm MM for even m, compare Proposition 32.1.

Finally the following simple fact shows some correlation between the classes of commutative quadratic and bilinear algorithms, compare also Remark 32.2.

Proposition 32.3 [W70]. Every commutative quadratic algorithm A for the evaluation of a set of bilinear forms $b_\gamma(X,Y) = q_\gamma(V)$, $\gamma = 0,1,\ldots,\Gamma-1$, can be transformed into a bilinear algorithm $\tilde{A}$ for the evaluation of the same set of bilinear forms such that

$$\rho(\tilde{A}) \leq 2\rho(A),\ as(\tilde{A}) \leq 2\ as(A),\ sm(\tilde{A}) \leq 2\ sm(A).$$

Proof. Represent the algorithm A by (32.3) and (32.4). Rewrite

$$L_h(V) = L_h(X) + L_h(Y),\ L_h'(V) = L_h'(X) + L_h'(Y)$$

where $L_h(X)$ and $L_h'(X)$ are linear forms in the x-variables; $L_h(Y)$ and $L_h'(Y)$ are linear forms in the y-variables. Substitute these expressions into (32.3) and obtain that

$$b_\gamma(X,Y) = \sum_{h=0}^{M-1} f''(\gamma,h)(L_h(X) + L_h(Y))(L_h'(X) + L_h'(Y)) =$$

$$\sum_{h=0}^{M-1} f''(\gamma,h)(L_h(X)L_h'(Y) + L_h(Y)L_h'(X) + L_h(X)L_h'(X) + L_h(Y)L_h'(Y))$$

for all γ.

It remains to note that we may delete the terms $L_h(X)L_h'(X)$ and $L_h(Y)L_h'(Y)$ for all h without any influence on the bilinear forms $b_\gamma(X,Y)$ that include no terms of the form $x_\alpha x_{\alpha'}$ or $y_\beta y_{\beta'}$. ∎

33. λ-algorithms for the Multiplication of Matrices of Small and Moderate Sizes.

As follows from Theorem 23.7, bilinear λ-algorithms can efficiently compute approximations to the products of pairs of matrices. In Example 4.1 we could see that the λ-rank of 2X2X3 MM is at most 10 while the rank of that problem is known to be 11, at least over the ring Z of integer constants, see [HK71]. Similarly bilinear λ-algorithms for (2r) X (2r) X (3r) MM whose degree is 8 and λ-ranks are $M = 4r^3+12r^2+8r$ for all r have been presented in [P79] and this is substantially less than the ranks of all known conventional bilinear algorithms for the same problems. In particular for r=10, M=5280, $3 \log M/\log(12r^3) < 2.7378$. Also for square MM of moderate sizes some λ-algorithms of degree 5 from [P79] substantially improve over the algorithms of Section 31. In particular for all n of the form

$n=6gh$ the λ-algorithms for $n \times n$ MM of degree 5 and of rank $M = n^3/3 + 3n^2 + 4(3g-2)h^2 + 16(4g-1)h$ have been presented in [P79]. This gives the following table of the values of M and log M/log n for some specific g and h, compare Table 31.2.

Table 33.1

g	h	n	M	log M/log n
1	4	24	6616	2.76813
2	2	24	6848	2.77897
4	1	24	6592	2.76699
1	5	30	12056	2.76295
5	1	30	12040	2.76255
1	6	36	19872	2.76183
2	3	36	19904	2.76228
3	2	36	19920	2.76250
6	1	36	19872	2.76183
1	7	42	30506	2.76260
7	1	42	30530	2.76281
1	8	48	44360	2.76403
2	4	48	44416	2.76435
4	2	48	44480	2.76473
8	1	48	44412	2.76433

The above bilinear λ-algorithms can be applied recursively. Then their degrees will grow much slower than the sizes of the problems they solve, see Proposition 6.2, and consequently the relative efficiency of the recursive λ-algorithms will rapidly grow with the size of the problem, compare Theorems 6.3 and 23.7. On the other hand, we are most interested in avoiding the recursion completely in order to be able to apply the algorithms to the multiplication of matrices of smaller sizes. For the latter problem, some commutative quadratic λ-algorithms (to be considered in this section) are even more efficient than bilinear λ-algorithms.

In order to define the class of commutative quadratic λ-algorithms, extend the class of conventional commutative quadratic algorithms following Remark 6.1 and combining it with Definition 32.1. The notation $\rho(A)$, as(A), sm(A) is also extended to every commutative quadratic λ-algorithm A where it is assumed that $\lambda = 2^{-G}$ for a sufficiently large natural G and that the terms of the higher degrees in λ have been deleted from the outputs.

Here is an example of a commutative quadratic λ-algorithm A of degree 1 for mX2X1 MM.

Example 33.1 [B84].

$$x_{i0}y_{00} + x_{i1}y_{10} = (\lambda^{-1}y_{10} + x_{i0})(y_{00} + \lambda x_{i1}) - \lambda^{-1}s_0(Y) - \lambda q_i(X) \qquad (33.1)$$

where $s_0(Y)$ and $q_i(X)$ are defined by (32.2), $i=0,1,\ldots,m-1$.

To define the desired λ-algorithm A, choose a sufficiently large natural G, let $\lambda = 2^{-G}$, and successively evaluate $\lambda^{-1}y_{10}$, $\lambda^{-1}s_0(Y) = (\lambda^{-1}y_{10})\, y_{00}$, $\lambda^{-1}y_{10} + x_{i0}$, $y_{00} + \lambda x_{i1}$, $(\lambda^{-1}y_{10} + x_{i0})(y_{00} + \lambda x_{i1})$, and finally $(\lambda^{-1}y_{10} + x_{i0})(y_{00} + \lambda x_{i1}) - \lambda^{-1}s_0(Y)$ for all i. The latter values define the desired approximation to the mX2X1 matrix product. ■

The above λ-algorithm A involves m+1 multiplications, 3m additions and subtractions, and m multiplications and 1 division by $\lambda = 2^{-G}$ (these multiplications and divisions amount to m+1 shifts of the radix point), that is,

$$\rho(A) = m+1,\ as(A) = 3m,\ sm = m+1. \qquad (33.2)$$

(This can be compared with 2m nonscalar multiplications and m additions in the straightforward algorithm. The latter bounds cannot be improved unless we simultaneously allow commutativity and approximation, see [B84].)

Propositions 32.1 and 32.3 are immediately extended to the case of commutative quadratic λ-algorithms. Applying the extension of Proposition 32.1, we can transform the λ-algorithm (33.1) into a commutative quadratic λ-algorithm A' of degree 1 for mX(2r)Xp MM for arbitrary m,p,r such that

$$\rho(A') = (m+1)pr,\ as(A') = (3r + 1)mp,\ sm = (m+1)pr. \qquad (33.3)$$

Next we will show that the latter estimates can be reduced by eliminating some redundant operations.

Example 33.2.

$$\sum_j x_{ij}y_{jk} = \sum_{h=0}^{r-1}(\lambda^{-1}y_{2h+1,k} + x_{i,2h})(y_{2h,k} + \lambda x_{i,2h+1}) - \lambda^{-1}s_k(Y) - \lambda q_i(X) \quad (33.4)$$

where $q_i(X)$ and $s_k(Y)$ are defined by (32.10), $i=0,1,\ldots,m-1$; $k=0,1,\ldots,p-1$.

It is assumed that the terms $\lambda q_i(X)$ should be deleted for all i,k and that $\lambda^{-1}s_k(Y)$ should be computed from the identity

$$\lambda^{-1}s_k(Y) = \sum_{h=0}^{r-1}(\lambda^{-1}y_{2h+1,k})y_{2h,k},\ k=0,1,\ldots,p-1.$$

Then it is easily verified that the resulting commutative quadratic λ-algorithm A of degree 1 approximately evaluates $m\times(2r)\times p$ matrix products and that

$$\rho(A) = (m+1)pr,\ as(A) = 3mpr + p(r-1),\ sm(A) = (m+p)r. \qquad (33.5)$$

This is a noticeable improvement over (33.3), compare also (32.11) and (33.2).

The dual λ-algorithm A^* for $p\times(2r)\times m$ MM also involves as many operations of each kind as A does, see (32.6).

In particular if $m=p=2r$, then (33.4) defines a commutative quadratic λ-algorithm A for $n\times n$ MM such that

$$\rho(A) = 0.5(n+1)n^2,\ as(A) = 1.5n^3 + 0.5n^2 - n,\ sm(A) = n^2, \qquad (33.6)$$

compare (32.12).

In the next tables we will present the numbers of operations involved in the λ-algorithm (33.4), in modified Strassen's algorithm, and in the straightforward algorithm for $n\times n$ MM for $n=6$ and $n=2^s$, $s=1,2,3,4,5$. We will use the following modification of Strassen's algorithm for 2X2 MM due to S. Winograd (see [BM], pp. 45-46), which involves only 7 multiplications and 15 additions and subtractions (at least 15 additions and subtractions are required in any bilinear algorithm of rank 7 for 2X2 MM, see [Pr73]). Actually every bilinear algorithm of rank 7 for 2X2 MM can be transformed into Strassen's by a simple linear transformation of inputs and outputs, see Theorem 3 in [P72] or see [dG].

<u>Example 33.3.</u>

$$L_0 = x_{10} + x_{11},\ L_1 = L_0 - x_{00},\ L_2 = x_{00} - x_{10},\ L_3 = x_{01} - L_1,$$

$$L'_0 = y_{01} - y_{00},\ L'_1 = y_{11} - L'_0,\ L'_2 = y_{11} - y_{01},\ L'_4 = L'_1 - y_{10},$$

$$q_0 = L_0L'_0,\ q_1 = L_1L'_1,\ q_2 = L_2L'_2,\ q_3 = L_3y_{11},\ q_4 = x_{11}L'_4,$$

$$q_5 = x_{00}y_{00},\ q_6 = x_{01}y_{10};\ t_0 = q_1 + q_5,\ t_1 = t_0 + q_2,\ t_2 = t_0 + q_0;$$

$$x_{00}y_{00} + x_{01}y_{10} = q_5 + q_6;\ x_{00}y_{01} + x_{01}y_{11} = t_2 + q_3;$$

$$x_{10}y_{00} + x_{11}y_{10} = t_1 - q_4;\ x_{10}y_{01} + x_{11}y_{11} = t_1 + q_0.$$

Table 33.2, λ-algorithm (33.4), compare (33.6)

n	rank	additions subtractions	shifts of radix point
2	6	13	4
4	40	102	12
6	126	339	36
8	288	796	64
16	2176	6264	256
32	16896	49648	1024

Table 33.3. Winograd's Modification of Strassen's Algorithm

n	rank	additions subtractions
2	7	15
4	49	165
8	343	1395
16	2401	10725
32	16807	78915

Table 33.4. Straightforward Algorithm

n	rank	additions, subtractions
2	8	4
4	64	48
6	216	180
8	512	448
16	4096	3840
32	32768	31744

We can see that the λ-algorithm (33.4) does not involve too many linear operations and that it is even superior over modified Strassen's algorithm by that criterion if the problem size is moderate. Note also that $n=32$ is the crossover point where the λ-algorithm (33.4) first involves slightly more nonscalar multiplications than modified Strassen's algorithm does.

For the sake of completeness, we recall that the λ-algorithm for 3X3 MM from [S81] has the following parameters, rank 21, 66$\pm$, and 51 shifts against rank 27 and 18$\pm$ in the straightforward algorithm. The latter algorithm certainly wins the competition. On the other hand, the λ-algorithms (33.4) may compete in the speed with the straightforward algorithm for nXn MM already for n=4 and n=6, compare Tables 33.2 and 33.4, and also in the case of rectangular MM, say, for 5X2X1 MM, see (33.2).

34. The Classes of Straight Line Arithmetical Algorithms and λ-algorithms and Their Reduction to Quadratic Ones.

In this section we will extend the previously considered classes of algorithms and λ-algorithms to the most general customary classes of the straight line arithmetical algorithms and λ-algorithms for the evaluation of a set of rational expressions (such a class of problems includes MM, SLE, Det, and MI). The only customary simplification in our definition is that we do not include branchings in those algorithms but actually our study will little change if we allow branchings, compare Remark 35.4 below. (See Section 40 and the end of Section 18 about further generalization of the computational scheme and compare [P80a].) In the present section we will show that the extension to the classes of arithmetical algorithms and λ-algorithms cannot help much in the acceleration of MM and more generally in the evaluation of a set of quadratic forms. Of course such an extension is required for the evaluation of higher degree polynomials; Example 34.1 at the end of this section will demonstrate the efficiency of the application of λ-algorithms to that problem.

Definition 34.1. Given a ring F and a set $\{Q_g(V),\ g=0,1,\ldots,G-1\}$ of rational expressions (quolynomials) in indeterminates $v_0,v_1,\ldots,v_{H-1}$ with coefficients from F. Then a sequence

$$p_s = R_s' \circ R_s,\ s=0,1,\ldots,S-1, \tag{34.1}$$

is called a straight line arithmetical algorithm A of an arithmetical complexity ar(A)=S for the evaluation of the set $\{Q_g(V)\}$ over the ring F if there exist natural $s(g) < S$, $g=0,1,\ldots,G-1$, such that

$$Q_g(V) = p_{s(g)} \text{ for all } g. \tag{34.2}$$

Here o designates an arithmetical operation (that is, addition, subtraction, multiplication, or division) and each R_s, R_s' for every s is either v_h, $0 \le h < H$, or p_q for $q < s$, or a constant from a ring F. Each operation s in (34.1) is called <u>linear</u> (<u>scalar</u>) if it is either an addition, or a subtraction, or a division by a constant, that is, $p_s = R_s'/R_s, R_s \in F$, or a multiplication by a constant, that is, $p_s = R_s' * R_s$ and $R_s \in F$ or $R_s' \in F$. Otherwise the operation is called <u>nonlinear</u> (<u>nonscalar</u>). As in the previous sections, $\rho(A)$; as(A); sm(A); ar(A) will designate the numbers of nonscalar (nonlinear) operations; additions and subtractions; linear (scalar) multiplications and divisions; total number of arithmetical operations involved in A, respectively.

The class of arithmetical algorithms includes the class of bilinear algorithms as its subclass where all nonlinear operations take the form of bilinear steps. (Check this for bilinear algorithms of our Examples 2.1,2.2 and for Strassen's algorithm for 2X2 MM.) Similarly the class of commutative quadratic algorithms is included here.

It is natural to ask if the divisions can be of any help in the cases where the outputs are polynomials (for instance, in the case of MM). The next example shows that divisions should not be simply discarded.

<u>Example 34.1</u>. The polynomial $Q(v) = v^{15}+v^{14}+\ldots+1$ can be rapidly evaluated as follows, $p_0 = v-1$, $p_1 = v*v$, $p_2 = p_1*p_1$, $p_3 = p_2*p_2$, $p_4 = p_3*p_3$, $p_5 = p_4-1$, $Q(v) = p_6 = p_5/p_0$. ■

On the other hand, divisions are not of much help in the case of MM because we have the following important result due to [St73].

<u>Proposition 34.1</u>. If a given set $\{Q_g, g=0,1,\ldots,G-1\}$ of <u>polynomials of degree at most 2</u> can be evaluated by a straight line arithmetical algorithm A over a field F of constants and if that algorithm A involves M nonlinear (nonscalar) operations, then the given set of polynomials can be evaluated by an arithmetical algorithm that also contains at most M nonlinear (nonscalar) operations and all of them are multiplications. Furthermore all constants of the latter algorithm belong to F if F is infinite or to a simple algebraic extension of F otherwise, compare Definition 5.1.

<u>Proof</u>. Let operation d be the first nonscalar division in the given algorithm A. Then p_s for s<H, R_d', R_d are polynomials. Represent them as the sums of homogeneous polynomials of degrees 0,1,2,... and drop all terms of degrees greater than 3. (That means computing modulo the <u>ideal</u> J <u>generated by</u> $\{v_i v_j v_k,\ 0 \le i \le j \le k < G\}$. The reduction modulo J of any polynomial of degree 2, such as the outputs Q_g, does not change them, of course.) Then the first nonlinear division can be simulated as follows (provided that $c_d \neq 0$, see below).

$$p_d \text{ modulo } J = (c_d' + l_d' + q_d')/(c_d + l_d + q_d) \text{ modulo } J = \tag{34.3}$$

$$(c_d' + l_d' + q_d')((1/c_d) - (1/c_d)^2(l_d + q_d) + (1/c_d)^3(l_d + q_d)^2) \text{ modulo } J.$$

Here c_d, l_d, q_d are the homogeneous polynomials of degrees 0,1, and 2 that represent the constant, the linear, and the quadratic parts of R_d, respectively, while c_d', l_d', q_d' similarly represent R_d'.

It is easy to verify that the nonlinear multiplications modulo J in (34.3) can be reduced to the single nonlinear multiplication of $(1/c_d)^2((c_d'/c_d)l_d - l_d')$ by l_d and to some linear operations. That way the first nonlinear division in the algorithm A has been replaced by a nonlinear multiplication. Perform similar transformation of the next nonscalar division and continue until no nonscalar divisions are left. It remains only to handle the case where the constant terms of some divisors equal 0. To avoid such a phenomenon, substitute the variables $v_h = u_h + \theta_h$, $h=0,1,\ldots,H-1$, where θ_h are constants such that $\tilde{c}_s \neq 0$ for all s, $s=0,1,\ldots,S-1$ (or for all s associated with the divisions of the algorithm), in the decompositions $R_s = \tilde{c}_s + \tilde{l}_s + \tilde{q}_s + \ldots$ of R_s as the sums of the homogeneous polynomials in the new variables $u_0, u_1, \ldots, u_{H-1}$. Such constants θ_h, $h=0,1,\ldots,H-1$, can be always chosen in F if F has sufficiently many elements (in particular, if F is infinite) or among the roots of an irreducible polynomial of sufficiently high degree over F otherwise. (Such a polynomial defines the desired simple algebraic extension of F, compare Definition 5.1.) ∎

We do not know any examples where the use of divisions or of terms of degrees higher than 2 would actually accelerate nXn MM for any specific n. Proposition 34.1 and the next simple fact show that, even if such an acceleration is possible, then there exists almost as fast algorithms for the same problem that do not involve divisions. (On the other hand, the use of divisions might facilitate the search for such fast algorithms.)

<u>Proposition 34.2</u>, see [W70]. Let a straight line arithmetical algorithm A over a ring F for a quadratic problem (that is, for the problem of the evaluation of a set of polynomials of degrees at most 2) involve M nonlinear operations, all of which are multiplications. Then there exists a commutative quadratic algorithm A' over F for the same quadratic problem such that

$$\rho(A') = \rho(A), \; as(A') \leq 3as(A) + 5\rho(A), \; sm(A') \leq 3sm(A) + 2\rho(A), \qquad (34.4)$$

and consequently

$$ar(A') \leq 9ar(A).$$

<u>Proof</u>. Examine the scheme (34.1) for the given algorithm A. Note that all p_s in (34.1) are polynomials in V because A involves no nonlinear divisions. As in the proof of Proposition 34.1, compute modulo J, J being the ideal generated by $\{v_i v_j v_k, \; 0 \leq i \leq j \leq k < G\}$, so that the terms of degrees higher than two are deleted and $p_s = c_s + l_s + q_s$ modulo J for all s. Here again c_s, l_s, q_s are the sums

of all constants, all linear, and all quadratic terms of p_s, respectively. Surely, the quadantic outputs do not change if the computation is modulo J.

Now every nonlinear operation of the algorithm takes the form

$$p_s = (c_g + l_g + q_g)(c_h + l_h + q_h) \text{ modulo } J.$$

Consequently

$$p_s = l_g l_h + c_g(c_h + l_h + q_h) + (l_g + q_g)c_h \text{ modulo } J,$$

compare (34.3). Therefore such an operation has been reduced modulo J to one nonlinear operations, that is, to the multiplication $l_g l_h$, and to seven linear operations. Thus we arrive at the desired commutative quadratic algorithm A' and immediately verify (34.4). ■

Remark 34.1. Combining Propositions 32.3, 34.1, and 34.2 immediately implies Theorem 2.4.

Finally, we we will define the class of straight line arithmetical λ-algorithms and note that our study in this section can be easily extended to that class.

Definition 34.2. Let $F[\lambda]$ be the ring of polynomials in λ over a field or a ring F. Let p_s, $s=0,1,\ldots,S-1$, be defined by (34.1) where the constants are from $F[\lambda]$. Let there exist natural $s(g) < S$ such that

$$\lambda^d \, Q_g(V) - p_{s(g)} = \lambda^{d+1} \, P_g(\lambda), \; g=0,1,\ldots,G-1, \tag{34.5}$$

where $P_g(\lambda)$ are polynomials in λ for all g, compare (34.2). Then the sequence (34.1) is called a straight line arithmetical λ-algorithm of degree d for the evaluations of a given set of rational functions (quolynomials) $\{Q_g(V), g=0,1,\ldots,G-1\}$ over the field or ring F.

Remark 34.2. The statements and the proofs of Propositions 34.1 and 34.2 are immediately extended to the case of straight line arithmetical λ-algorithms. Proposition 32.1 can be also easily extended to the classes of straight line arithmetical algorithms and λ-algorithms.

Remark 34.3. If we substitute the field $F(\lambda)$ of formal power series in λ for the ring $F[\lambda]$ of polynomials in λ in Definition 34.2, then we arrive at an equivalent version of Definition 34.2 by replacing (34.5) by the following requirement,

$$Q_g(V) - p_{s(g)} = \lambda \, P_g(\lambda), \; g=0,1,\ldots,G-1,$$

compare Remark 6.1.

Finally, here is an example of an arithmetical λ-algorithm for the evaluation of a polynomial $z(\underline{v};x) = \sum_{i=0}^{n} v_i x^i$ at a point x. That λ-algorithm demonstrates the power of the class of algorithms introduced in Definition 34.2 and should counter rather negative character of the results of this section. The example is taken from [B84] where it is derived from commutative quadratic λ-algorithms for matrix-vector multiplication. We will give only the explicit representation of the final λ-algorithm, which reminds us of the well-known schemes for the evaluation of polynomials with preconditioning, see [K].

Example 34.2. (Evaluation of a polynomial $z(\underline{v};x) = \sum_{i=0}^{n} v_i x^i$, n is even.)

$$r=xx,\ s=rx,\ z_1=(\lambda v_n+x)(v_{n-1}+\lambda^{-1}r) + v_{n-2}-\lambda^{-1}s,\ z_{i+1}=(\lambda z_i+x)(v_{n-2i-1}+\lambda^{-1}r)+v_{n-2i-2}-\lambda^{-1}s,$$

$i=1,2,\ldots,0.5n-1$, $z(\underline{v},x) = z_n$. ∎

The latter algorithm involves $0.5n+2$ nonscalar multiplications, $2n$ additions and subtractions and $0.5n+2$ multiplications and divisions by λ, which can be reduced to the shifts of the radix point by choosing $\lambda=2^{-G}$ for natural G. This is to be compared with n multiplications and n additions involved in Horner's rule, $z_0=a_n$, $z_i=z_{i-1}x+a_{n-i}$, $i=1,2,\ldots,n$. (Horner's rule is the optimum method in the class of arithmetical algorithms for the exact evaluation of n-th degree polynomials at a point, see Theorems 37.1 and 37.2 in Section 37.)

It seems promising to try to construct efficient arithmetical λ-algortihms for the problems SLE, MI, and Det.

35. The Basic Active Substitution Argument and Lower Bounds on the Ranks of Arithmetical Algorithms for Matrix Multiplication.

In this and in the next two sections we will recall the best current lower bounds on the arithmetical complexity of mXnXp MM. The bounds rely on the techniques of basic active substitutions, see Remark 35.5 and the proofs of Theorem 37.1 and 37.2 below; the bounds are linear in the number of inputs and outputs, so that they remain much less than the current asymptotically best upper bounds of the order $n^{2.495\ldots}$ in the case of nXn MM.

In this section we will establish the lower bounds on the ranks of conventional arithmetical algorithms for MM. In the next section we will estimate the ranks of λ-algorithms for MM. Then we will study the additive complexity of MM and will use that occasion to introduce a combinatorial concept that characterizes the logical complexity (asynchronicity) of bilinear and linear computations. In the process of that study we will show how the power of the basic active substitution argument can be enhanced by combining it with Propositions 34.1 and 34.2, with the duality Theorem 2.3, and with the properties of the representation of quadratic algorithms by matrix-vector equations.

We will start with the following result.

Theorem 35.1. For all m,n,p and for all fields F that contain sufficiently many distinct elements, every straight line arithmetical algorithm for mXnXp MM over F involves at least $n(m+p-1)+m-1$ nonscalar operations if $p>1$ and at least mn nonscalar operations if $p=1$.

Definition 35.1. The commutative rank of mXnXp MM over a ring F of constants, $\varphi_F(m,n,p)$, is defined as the minimum rank $\rho(A)$ in all commutative quadratic algorithms A for mXnXp MM over F. Hereafter $\varphi(m,n,p)$ will stand for $\varphi_F(m,n,p)$ in those relations that hold for all fields F.

Remark 35.1. Theorem 35.1 immediately follows from the next two relations,

$$\varphi(m,n,1) = mn, \tag{35.1}$$

$$\varphi(m,n,p) \geq n(m+p-1)+m-1 \text{ if } p > 1, \tag{35.2}$$

and from Propositions 34.1 and 34.2. The straightforward algorithm for MM implies that the bound (35.1) is sharp.

Remark 35.2. The bound (35.1) is due to [W70]. The bound (35.2) is due to [LW] and was preceeded by the bound

$$\rho(m,n,p) \geq n(m+p-1)+m-1 \text{ for } p > 1$$

due to [BD78].

Proof of (35.1). Recall (32.3),(32.4). In our case the v-variables are x_{ij} and y_{jk} for all i,j,k. Let $v_0 = x_{00}$. (Otherwise rename the variables v_σ.) Let $f(0,0) \neq 0$. (Otherwise rearrange the order of L_q and interchange L_q and L'_q if needed.) Then

$$L_0(V) = f(0,0)x_{00} + L_0^*(V)$$

where $L_0^*(V)$ is an x_{00}-free linear form in the v-variables. Substitute

$$x_{00} = -(f(0,0))^{-1} L_0^*(V) \tag{35.3}$$

into (32.3),(32.4). Note that this turns $L_0(V)$ into 0, that is, turns the bilinear multiplication $L_0(V)L'_0(V)$ into the passive operation $0*L'_0(V)$, and reduces the number of free variables (indeterminates) by 1. The substitution (35.3) can be interpreted also as the inclusion of $L_0(V)$ into the ring of constants. Repeat the same procedure and turn $L_{h(s)}(V)$ or $L'_{h(s)}(V)$ into zeroes (include one of those two linear forms into the ring of constants) for mn distinct h(s) by the successive substitutions of linear forms for the x-variables $x_{i(s)j(s)}$, $s=0,1,\dots,mn-1$. The process will include at least mn steps because fewer than mn linear substitutions of the x-variables into the mXnX1 product XY would leave at least one bilinear term in XY. (The latter fact is very easy to check if the linear substitutions are applied to the entries of the matrix X in the representation (35.5) below.) Each step reduces the number of nonlinear operations in (32.3) by at least 1. (Historically the name active came from [P62] where nonscalar multiplications eliminated by the substitutions were called active; the motivation for such a name will be also seen in the proofs of Theorems 37.1 and 37.2 in Section 37.) ■

Surely the above argument can be applied to the algorithm (32.3),(32.4) for mXnXp MM for all m,n,p. Furthermore in that case we may rather easily increase the lower bound to

$$\varphi(m,n,p) \geq n(m+p-1). \tag{35.4}$$

Indeed, eliminate n(m-1) nonlinear steps of (32.3) by appropriate substitutions of linear forms for the x-variables x_{ij} for $i=0,1,\ldots,m-2$; $j=0,1,\ldots,n-1$, as in the above proof of (35.1). This time, however, preserve the variables $x_{m-1,j}$ for all j and eliminate np more nonlinear steps of (32.3) by appropriate subsequent substitutions of linear forms for all of the y-variables y_{jk}.

We could get a further extension of (35.4) to (35.2) if we derive the same lower bound (m+p-1)n on the rank of a smaller problem. We may note that even if we exclude up to m-1 output entries from one of the columns of XY, then still the remaining outputs will depend on all of the mn+np entries of X and Y. Thus we may expect that the previous argument or its modification will give the lower bound (m+p-1)n on the rank of the algorithms for the evaluation of the remaining outputs. Then we may expect to increase that lower bound by m-1 if we require to compute the deleted m-1 outputs.

To get a formal proof, we will follow [LW] and will rely on the representation of mXnXp MM in the following form of matrix-vector equation, compare [W70], [F71],[F72], [St72]. Represent the matrices Y and Z=XY as the column-vectors $\underline{y} = [y_{00},y_{10},\ldots,y_{n-1,0},y_{01},y_{11},\ldots,y_{n-1,p-1}]^T$ and $\underline{z} = [z_{00},z_{01},\ldots,z_{m-1,0},z_{10},z_{11},\ldots,z_{m-1,p-1}]^T$, respectively. Then the matrix product Z=XY can be rewritten as the matrix-vector product

$$\underline{z} = D(X)\underline{y} \tag{35.5}$$

where D(X) is the mpXnp block-diagonal matrix whose all p diagonal blocks coincide with the matrix X.

The identities (32.3) are now represented by the matrix-vector equation,

$$D(X)\underline{y} = G\,\underline{K} \tag{35.6}$$

where $\underline{K} = [K_0, K_1, \ldots, K_{M-1}]^T$, $K_h = L_h(V)L_h'(V)$, $\underline{K}$ is the column-vector of the M products $L_h(V)L_h'(V)$, and G is the mpXM matrix of the coefficients $f''(\gamma,h)$ in (32.3). Let (35.6) be the representation of the algorithm (32.3) of the rank $\varphi(m,n,p)$ for mXnXp MM.

Since all entries of $\underline{z} = D(X)\underline{y}$ are linearly independent over F, the rows of D(X) and of G are also linearly independent. Then there should be an mXn nonsingular square submatrix in the first m rows of G. We may assume that such a submatrix $\tilde{G}$ lies in the northwestern corner of G (interchanging the columns of G otherwise). Let $H = \tilde{G}^{-1}$ and $B = \mathrm{diag}\,\{H,H,\ldots,H\}$ be the mpXmp block-diagonal matrix that consists of exactly p diagonal blocks H. Multiply the identity (35.6) by B and obtain the following identity.

$$D(HX)\underline{y} = (BG)\underline{K}. \qquad (35.7)$$

We may consider $\tilde{X} = HX$ the matrix of input-variables whose entries are indeterminates, so that the matrix-vector equation

$$\underline{\tilde{z}} = D(\tilde{X})\underline{y}, \ \underline{\tilde{z}} = B\underline{z}, \ \tilde{X} = HX,$$

again represents mXnXp MM, compare (35.5). Then (35.7) defines a commutative quadratic algorithm for mXnXp MM such that the mXm identity matrix I stands in the northwestern corner of the matrix BG. To simplify the notation, we will assume that the latter property holds already for the original matrix G of (35.6).

Next represent the first m columns of G as the pmXm matrix $[G_0, G_1, \ldots, G_{p-1}]^T$ where G_k is the mXn submatrix of G in the intersection of the first m columns of G with its rows km+i, i=0,1,...,m-1.

Since $G_0 = I$, we may turn the blocks of G represented by G_k, k=1,2,...,m-1, into the null matrices by premultiplying G by the following matrix,

$$C = \begin{bmatrix} I & 0 & \cdots & 0 \\ -G_1 & I & \cdots & 0 \\ -G_2 & 0 & \cdots & 0 \\ \cdot & \cdot & \cdots & \cdot \\ \cdot & \cdot & \cdots & \cdot \\ \cdot & \cdot & \cdots & \cdot \\ -G_{p-1} & 0 & \cdots & I \end{bmatrix}.$$

Premultiply both sides of (35.6) by C and obtain the following matrix-vector equation,

$$CD(X)\underline{y} = (CG)\underline{K},$$

which represents a bilinear algorithm of the same rank $\varphi(m,n,p)$ for computing

$$C\underline{z} = CD(X)\underline{y}. \qquad (35.8)$$

Let us delete the first m-1 identities of (35.8). This reduces the size of the problem and the rank of the algorithm by m-1. The resulting subproblem of the rank $\varphi(m,n,p)-m+1$ takes the following form,

$$\underline{z}^* = A(X)\underline{y}. \qquad (35.9)$$

Here the vector $\underline{z}^*$ is obtained from the vector $C\underline{z}$ by deleting its first m-1 entries; the $(pm-m+1)Xm$ matrix $A(X)$ is obtained from the mpXm matrix $CD(X)$ by deleting its first m-1 rows, so that

$$A(X) = \begin{bmatrix} \underline{x}_{m-1} & 0 & 0 & \cdots & 0 \\ -G_1X & X & 0 & \cdots & 0 \\ -G_2X & 0 & X & \cdots & 0 \\ \vdots & & & & \\ -G_{p-1}X & 0 & 0 & \cdots & X \end{bmatrix}$$

where $\underline{x}_{m-1} = [x_{m-1,0}, x_{m-1,1}, \ldots, x_{m-1,n-1}]$.

Now we may apply the active substitution argument to the subproblem (35.9) of (35.8) and prove that the rank of that subproblem is not less than $(m+p-1)n$. Indeed, apply the successive substitutions of the same form as (35.3), that is,

$$v_\sigma = -L_q^* \,/\, f^*(\sigma,q) \qquad (35.10)$$

such that $f^*(\sigma,q) \neq 0$, L_q^* does not depend on v_σ, $K_q = L_qL_q'$ (see (35.6)), and either L_q, or L_q', or both of them are equal to $L_q^* + f^*(\sigma,q)v_\sigma$. We first successively choose all of the $(m-1)p$ variables x_{ij} as v_σ for all j and for $i<m-1$. (There must exist q such that $f^*(\sigma,q) \neq 0$ in this case due to the bilinear dependence of the outputs $\underline{z}^*$ on x_{ij} for all i,j and on the y-variables.) The remaining bilinear problem

$$\underline{\tilde{z}} = \tilde{A}(X,Y)\underline{y}$$

is defined by the (mp-m+1) X mp matrix $\tilde{A}(X,Y)$ where deleting all rows but rows 1+km, k=0,1,...,p-1, we obtain the following lower triangular block-submatrix,

$$\begin{bmatrix} \underline{x}_{m-1} & \underline{0} & \underline{0} & \underline{0} & \cdots & \underline{0} \\ \cdot & \underline{x}_{m-1} & \underline{0} & \underline{0} & \cdots & \underline{0} \\ \cdot & \cdot & \underline{x}_{m-1} & \underline{0} & \cdots & \underline{0} \\ \cdot & \cdot & \cdot & \cdot & \cdots & \cdot \\ \cdot & \cdot & \cdot & \cdot & & \underline{x}_{m-1} \end{bmatrix}.$$

All superdiagonal blocks of that matrix are filled with the null vectors $\underline{0}$. The next np linear substitutions will eliminate at least np nonscalar operations until all of the np variables y_{jk} have been eliminated. Then the total reduction of the rank $\varphi\rho(m,n,p)$ of the original problem will be m-1+n(m-1)+np = n(m+p-1)+m-1. This proves (35.2). ■

Combining Theorem 35.1 with Theorem 2.3 and Proposition 32.1 we obtain the following corollary.

Corollary 35.2.

$$\rho(1,n,p) = \varphi\rho(1,n,p) = np, \ \rho(m,1,p) = mp, \ \varphi\rho(m,n,1) = \rho(m,n,1) = mn.$$

$$\rho(m,n,p) \geq \varphi\rho(m,n,p) \geq n(m+p-1)-1 + \text{maximum}\{m,p\} \text{ if } m>1,\ p>1.$$

$$\rho(n,n,n) \geq \varphi\rho(n,n,n) \geq 2n^2-1.$$

$$\rho(m,n,p) \geq m(n+p-1)-1+\text{maximum}\{n,p\} \text{ if } n>1,\ p>1.$$

One may ask if the latter lower bound on $\rho(m,n,p)$ can be extended also to the case of $\varphi\rho(m,n,p)$. Such an extension has not been proven yet. Let us derive a little weaker estimates, that is,

$$\varphi\rho(m,n,p) \geq m(n+p-1), \ \varphi\rho(m,n,p) \geq p(m+n-1). \tag{35.11}$$

Remark 35.3. The two lower bounds of (35.11) imply each other, due to Proposition 32.1.

Proof of (35.11), compare [BM, pp. 24-25]. Let the bound $\varphi\rho(m,n,p)$ be reached in an algorithm (35.6). Perform successive active substitutions of the variables x_{ij} for all i and for j=1,2,...,n-1 such that the rank has been reduced by m(n-1). The resulting algorithm will be represented by the matrix-vector equation of the form

$$A^*(X,Y)\underline{y} = G^*\underline{K}^*$$

where $A^*(X,Y)$ has a block-triangular submatrix of the size mpXp whose p diagonal blocks are equal to the same column-vector $[x_{i0}, i=0,1,\dots,m-1]$. Then there still must exist mp linearly independent expressions among the outputs of the associated algorithm. Their existence implies the lower bound mp on the number of remaining linearly independent products $L_q(V)\ L_q'(V)$, that is, the rank is still not less than mp after its reduction by m(n-1). ■

Due to Propositions 34.1 and 34.2, we may restate the above lower bounds on $\varphi(m,n,p)$ as follows.

Theorem 35.3. For all m,n,p and for all fields F that contain sufficiently many distinct elements, every straight line arithmetical algorithm A for mXnXp MM over F involves at least ar(A) nonscalar operations such that

$$\rho(A) \geq \text{maximum}\{n(m+p-1)+m-1, n(m+p-1)+p-1, m(n+p-1), p(m+n-1)\} \text{ if } m>1,\ p>1,$$

$$\rho(A) \geq mn \text{ if } p=1,\ \rho(A) \geq np \text{ if } m=1.$$

The lower bounds of Corollary 35.2 on the rank $\rho(m,n,p)$ are not sharp, at least over the field Z(2) of integers modulo 2, as follows from the next two resutls. (These results are trivially extended from the computation over Z(2) to the computation over the ring of integers Z, see (5.6). We will omit their proofs and some of their straightforward extensions that can be obtained via the duality Theorem 2.3.)

Theorem 35.4, [HK69],[HK71].

$$\rho_{Z(2)}(2,2,3) = 11,\ \rho_{Z(2)}(3,3,2) = 15.$$

Theorem 35.5, [JJT]. Let m>2, p>1. Then

$$\rho_{Z(2)}(m,n,p) \geq n(m+p-2) + \lceil 36n/23 \rceil + \lceil m/2 \rceil - 2.$$

In particular,

$$\rho_{Z(2)}(n,n,n) \geq 2n^2 - 2n + \lceil 36n/23 \rceil + \lceil n/2 \rceil - 2 \geq 2n^2 + (3n/46) - 2 \text{ for } n>2.$$

Remark 35.4. Retracing the steps of the above proofs of the lower bounds, we can see that those bounds can be applied to all specific input matrices X and Y except for ones where some linear equations in the x-variables and in the y-variables hold (such as (35.3) and (35.10)). The set V^* of all such inputs is the union of the linear subspaces of lower dimensions in the (mn+np)-dimensional linear space of all input variables. If we restrict the total number of arithmetical operations in the algorithms by some finite upper bound and require that F be an infinite ring but that all constants belong to its fixed finite subset, then the set V^* will be the union of finitely many linear subspaces of lower dimensions. Similarly we conclude that our lower bounds will not change if we allow finitely many branchings in arithmetical algorithms. This remark can be applied also to all lower bounds to be derived in the next sections. For that reason we will write "arithmetical algorithm" rather than "straight line arithmetical algorithm" hereafter.

Remark 35.5. The basic active substitution techniques will be further demonstrated in Sections 37. Those techniques and their extensions remain the most effective tools for establishing lower bounds on the arithmetical complexity of MM and of several other arithmetical computational problems, compare [St72], [BM], [AS]. In particular, in [AS] those techniques are nontrivially extended (via the classical structure theory of Wedderburn) to the unified proof of the lower bounds on the ranks of polynomial and matrix multiplication and Disjoint MM. Particularly the sharp bound $\rho(S \otimes (2,2,2)) = 7S$ has been proven in [AS]. Actually there are only two general classes of lower bounds on the arithmetical complexity in addition to the bounds obtained by the basic substitutions, that is, the linear lower bounds based on algebraic independence, see [Be], [BM], [K], and the nonlinear bounds based on the degree method due to V. Strassen, see [St73a], [BM], [Schn]. (The ingenious method due to M. Ben-Or, see [BO], works for both algebraic and combinatorial computational problems but it is close to Strassen's degree method in the case of arithmetical computations.) The degree method leads to nonlinear lower bounds on the complexity of interpolation, of computing elementary symmetric functions and continued fractions, see [St81], but helps neither for bilinear problems, such as MM, nor for Det, SLE, and MI.

36. Lower Bounds on the λ-rank and on the Commutative λ-rank of Matrix Multiplication.

In this section we will study the lower bounds on the λ-rank of λ-algorithms. Hereafter we will use the following notation.

Definition 36.1. $b\rho_F(m,n,p)$ and $cb\rho_F(m,n,p)$ will designate the minimum λ-ranks in all bilinear λ-algorithms and in all commutative quadratic λ-algorithms for mXnXp MM over a ring F of constants, respectively. $b\rho_F(m,n,p)$ and $cb\rho_F(m,n,p)$ will be called λ-rank and commutative λ-rank of mXnXp MM over F, respectively. (In [B84] they are called border rank and commutative border rank, respectively.) Hereafter we will substitute $b\rho(m,n,p)$ and $cb\rho(m,n,p)$ for $b\rho_F(m,n,p)$ and $cb\rho_F(m,n,p)$ in those relations that hold for all fields F. Similarly we will define rank, commutative rank, λ-rank, and commutative λ-rank of a bilinear problem $\{b_y(X,Y)\}$ and will designate them $\rho(\{b_y(X,Y)\})$, $c\rho(\{b_y(X,Y)\})$, $b\rho(\{b_y(X,Y)\})$, and $cb\rho(\{b_y(X,Y)\})$, respectively.

Let us derive some simple lower bounds on $cb\rho(m,n,p)$ and $b\rho(m,n,p)$. At first let us prove (using the linear independence argument) that

$$cb\rho(m,n,p) \geq mp \text{ for all } m,n,p. \tag{36.1}$$

Represent a commutative quadratic λ-algorithm of degree d and of λ-rank $M = cb\rho(m,n,p)$ for mXnXp MM as the set of the mp identities

$$\lambda^d \sum_j x_{ij}y_{jk} = \sum_{q=0}^{M-1} f''(k,i,q,\lambda)\, L_q(V,\lambda)\, L'_q(V,\lambda) \text{ modulo } \lambda^{d+1}$$

for all i,j and note that the mp left sides are linearly independent modulo λ^{d+1}, which means that there are at least mp different (and even linearly independent) products $L_q(V,\lambda)\, L'_q(V,\lambda)$. This immediately implies (36.1). ∎

Since $b\rho(m,n,p) \geq cb\rho(m,n,p) \geq mp$, Theorem 2.3 implies that

$$b\rho(m,n,p) \geq \text{maximum}\{mn,np,pm\} \text{ for all } m,n,p. \tag{36.2}$$

Similarly to (35.1) and (36.2) we deduce the following generalization of (36.1) and (36.2) (recall that Propositions 34.1 and 34.2 can be extended to the case of arithmetical λ-algorithms).

Theorem 36.1. For an arbitrary bilinear computational problem $\{b_y(X,Y)\}$, any bilinear λ-algorithm over a field F that contains sufficiently many distinct elements involves at least maximum$\{|X|,|Y|,\Gamma\}$ bilinear operations and any arithmetical λ-algorithm over F involves at least Γ nonscalar operations provided that |X|, |Y|, and Γ designate the numbers of x-indeterminates, of y-indeterminates, and of outputs that are linearly independent over F, respectively.

Applying Proposition 32.3 extended to the case of λ-algorithms we immediately deduce from (36.2) that

$$cb\rho(m,n,p) \geq 0.5 \text{ maximum}\{mn,np\} \text{ for all } m,n,p.$$

D. Bini proves the following stronger bound in [B84].

Theorem 36.2. $cb\rho(m,n,p) \geq 0.5n(m+p)$ for all m,n,p. (Consequently at least $0.5n(m+p)$ nonscalar operations are involved in any arithmetical algorithm for mXnXp MM.)

The proof relies on an auxiliary result, see (36.10) below, which is just a different version of Proposition 32.3. Proposition 32.3 can be rewritten as the following inequality, compare Definition 36.1,

$$2c\rho(\{b_\gamma(X,Y)\}) \geq \rho(\{b_\gamma(X,Y)\}) \text{ for an arbitrary bilinear set } \{b_\gamma(X,Y)\}. \quad (36.3)$$

Now we will represent the set of bilinear forms $\{b_\gamma(X,Y)\}$ by the set $\{A_\gamma\}$ of the matrices of their coefficients, so that

$$b_\gamma(X,Y) = \underline{x}^T A_\gamma \underline{y} \text{ for all } \gamma. \quad (36.4)$$

Here $\underline{x}$ and $\underline{y}$ are the column-vectors of the indeterminates. Every bilinear algorithm (2.5) for the problem $\{b_\gamma(X,Y)\}$ can be represented in the form of a set of decompositions of the matrices A_γ. Namely, recall (2.5) and represent the linear forms $L_q = L_q(X)$ and $L_q' = L_q'(Y)$ as follows,

$$L_q = \underline{x}^T \underline{f}(q), \; L_q' = (\underline{f}'(q))^T \underline{y} \text{ for all } q. \quad (36.5)$$

Here $\underline{f}(q)$ and $\underline{f}'(q)$ are the column-vectors of the coefficients of the linear forms L_q and L_q', respectively. Substitute (36.4) and (36.5) into (2.5) and deduce that

$$A_\gamma = \sum_{q=0}^{M-1} f''(\gamma,q)\underline{f}(q)(\underline{f}'(q))^T \text{ for all } \gamma. \quad (36.6)$$

Thus $\rho(\{b_\gamma(X,Y)\})$ is the minimum M in all decompositions (36.6). The set of matrices $\{A_\gamma\}$ can be interpreted as the set of the slabs of the 3-dimensional tensor t of the coefficients of the bilinear forms $\{b_\gamma(X,Y)\}$. Then the concepts of rank, λ-rank, commutative rank, and commutative λ-rank of the tensor t are defined as the same concepts applied to the bilinear set $\{b_\gamma(X,Y)\}$, that is, $\rho(t) = \rho(\{b_\gamma(X,Y)\})$, $b\rho(t) = b\rho(\{b_\gamma(X,Y)\})$, $c\rho(t) = c\rho(\{b_\gamma(X,Y)\})$, $cb\rho(t) = cb\rho(\{b_\gamma(X,Y)\})$. (36.3) relates $\rho(t)$ and $c\rho(t)$ but we need to extend that relation. We recall the definition of a commutative quadratic algorithm (32.3),(32.4) and assume that such an algorithm is applied to the evaluation of a bilinear set $\{b_\gamma(X,Y)\} = \{q_\gamma(V)\}$ where $V = \{X,Y\}$ designates the set of variables. Then we extend (36.5) as follows,

$$L_g(V) = [\underline{x}^T, \underline{y}^T] \begin{bmatrix} \underline{f}_x(g) \\ \underline{f}_y(g) \end{bmatrix}, \quad L_g'(V) = [(\underline{f}_x(g))^T, (\underline{f}_y'(g))^T] \begin{bmatrix} \underline{x} \\ \underline{y} \end{bmatrix} \tag{36.7}$$

Here $\underline{f}_x(g)$, $\underline{f}_y(g)$, $\underline{f}_x'(g)$, $\underline{f}_y'(g)$ are the coefficient vectors of the x-terms and of the y-terms of the linear forms $L_g(V)$ and $L_g'(V)$, respectively. Substitute (36.4) and (36.7) into a commutative quadratic algorithm (32.3),(32.4) and obtain the equivalent representation of that algorithm in the form of the following matrix equations,

$$\sum_g f''(\gamma,g)\, \underline{f}_x(g)\, (\underline{f}_x'(g))^T = \sum_g f''(\gamma,g)\, \underline{f}_y(g)\, (\underline{f}_y'(g))^T = 0, \tag{36.8}$$

$$\sum_g f''(\gamma,g)\, (\underline{f}_x(g)\, (\underline{f}_y'(g))^T + \underline{f}_y(g)\, (\underline{f}_x'(g))^T) = A_\gamma \text{ for all } \gamma.$$

(36.8) can be also interpreted as a decomposition (36.6) for a tensor $\tilde{t} + w - w^T$ of the size $(G+H, G+H, \Gamma)$ provided that the matrices A_γ are of the size $G \times H$. Here $\tilde{t}$ is the tensor represented by its $(G+H) \times (G+H)$ slabs as follows,

$$\tilde{t} = \{\begin{bmatrix} 0 & A_\gamma \\ 0 & 0 \end{bmatrix}, \gamma = 0,1,\dots,\Gamma-1\}, \tag{36.9}$$

w and w^T are the two tensors (of the same size) whose slabs are $\begin{bmatrix} 0 & 0 \\ W_\gamma^T & 0 \end{bmatrix}$ and $\begin{bmatrix} 0 & W_\gamma \\ 0 & 0 \end{bmatrix}$, respectively; $W_\gamma = \sum_g f''(\gamma,g)\, \underline{f}_y(g)\, (\underline{f}_x'(g))^T$, $\gamma = 0,1,\dots,\Gamma-1$. Therefore

$$\varphi(\{b_\gamma(X,Y)\}) = \varphi(t) \geq \rho(\tilde{t}+w-w^T),$$

which is the desired extension of (36.3), compare [How]. Note that $u=w-w^T$ is a skew-symmetric tensor of the size $(G+H) \times (G+H) \times \Gamma$, that is, such that all of the Γ slabs of that tensor are $(G+H) \times (G+H)$ matrices U_γ such that $U_\gamma^T = -U_\gamma$, $\gamma = 0,1,\dots,\Gamma-1$. Similarly we deduce that

$$cb\rho(t) \geq b\rho(\tilde{t} + u) \tag{36.10}$$

for some skew-symmetric tensor u. Let us deduce from (36.10) that

$$cb\,\rho(t) \geq 0.5\, b\,\rho(\tilde{t} + \tilde{t}^T) \tag{36.11}$$

where the tensor $\tilde{t}$ is defined by (36.9) and $\tilde{t}^T = \{\begin{bmatrix} 0 & 0 \\ A_y^T & 0 \end{bmatrix}\}$. ($\tilde{t}^T$ is obtained from $\tilde{t}$ by transposing coordinates 1 and 2.) Indeed, $b\rho(t) = b\rho(t^T)$ for every tensor t. In particular, $b\rho(\tilde{t} + u) = b\rho(\tilde{t}^T - u)$ if $u^T = -u$. On the other hand, $b\rho(u+v) \leq b\rho(u) + b\rho(v)$ for all tensors u and v. Therefore,

$$b\rho(\tilde{t} + \tilde{t}^T) = b\rho(\tilde{t} + u + \tilde{t}^T - u) \leq b\rho(\tilde{t} + u) + b\rho(\tilde{t}^T - u) \leq 2b\rho(\tilde{t} + u).$$

It remains to apply (36.10) and to obtain (36.11). ∎

Recall that the tensor $\tilde{t} + \tilde{t}^T$ is represented by the disjoint slabs $\begin{bmatrix} 0 & A_y \\ A_y^T & 0 \end{bmatrix}$ so that Theorem 36.1 implies that

$$b\rho(\tilde{t} + \tilde{t}^T) \geq |X| + |Y|$$

where $|X|$ and $|Y|$ are the numbers of the x-indeterminates and of the y-indeterminates of the bilinear set $\{b_y(X,Y)\}$ associated with the tensor $t = t(\{b_y(X,Y)\})$. We arrive at the following estimate,

$$cb\rho(\{b_y(X,Y)\}) \geq 0.5\,(|X| + |Y|). \tag{36.12}$$

In particular $|X|=mn$, $|Y|=np$ if $\{b_y(X,Y)\} = (m,n,p)$. This implies Theorem 36.2. ∎

Remark 36.1. Note that Theorem 36.1 and the bound (36.12) can be applied to Disjoint MM.

The lower bounds of Theorem 36.1 are sharp in the cases where m=1, n=1, or p=1; also the lower bound (36.1) is sharp if n=1, compare the straightforward algorithms. The lower bounds of Theorem 36.2 are sharp if m=1 or p=1, see Examples 33.1 and 33.2 and compare (32.6), (33.2), (33.5). For stronger lower bounds in the general case see [B84]. We will cite the following result omitting its proof that involves a rather elaborate analysis of tensor ranks.

Theorem 36.3 [B84]. For all natural n and for all infinite algebraically closed fields F,

$$b\rho_F(n,n,n) \geq 1.25\,n^2 + n - 1.25, \quad cb\rho_F(n,n,n) \geq 1.25\,n^2 + 0.5\,n - 1.25.$$

37. Basic Active Substitution Argument and Lower Bounds on the Number of Additions and Subtractions.

Our next objective is to establish some lower bounds on $as(m,n,p) = \min_A as(A)$, that is, on the minimum number of additions and subtractions involved in all arithmetical algorithms A for mXnXp MM. Unlike the number $\rho(A)$ of nonscalar operations, as(A) may blow up in the result of the reduction of A to a quadratic algorithm for mXnXp MM. Fortunately for the basic active substitution argument, it works not only for quadratic algorithms but also for all arithmetical algorithms and leads to the lower bounds on both numbers $\rho(A)$ and as(A).

Let us start with the demonstration by proving the following bound.

Theorem 37.1, [P62]. Let A be an arithmetical algorithm for the evaluation of a polynomial $z(\underline{v};x) = \sum_{i=0}^{n} v_i x^i$ of degree n over a field of constants F that contains sufficiently many elements. Then at least n nonlinear operations are involved in A.

Proof. Let A be represented by the sequence (34.1). Let operation q of that sequence be the first active nonlinear operation of A, that is, let there exist rational functions g(s,x) in x over F that do not depend on the vector $\underline{v}$ of the v-variables such that $p_s - g(s,x)$ are linear forms in $\underline{v}$ over F for s=0,1,...,q-1 but let such a function g(s,x) not exist for s=q. Then

$$R_q = \sum_{i=0}^{n} f(i,q)\, v_i + g^*(q,x)$$

where $g^*(q,x)$ is in the extension F(x) of the field F by the indeterminate x, f(i,q) are in F for all i, and not all of f(i,q) equal 0. Let, for instance, $f(n,q) \neq 0$. Then impose the following linear equation on the variable v_n,

$$v_n = -(f(n,q))^{-1} \sum_{i=0}^{n-1} f(i,q) v_i + g(x). \qquad (37.1)$$

where g(x) is an element of F(x) such that imposing (37.1) introduces no divisions by the identical 0 in (34.1). The substitution (37.1) can be also interpreted as adjoining R_q to the field F(x). The key idea of the substitution (37.1) is that such a substitution makes the operation q linear over F(x), that is, passive. Then consider the first remaining active nonlinear operation of A and turn it into linear one by imposing a linear equation of the same form as (37.1) and so on until no active nonlinear operations are left in the algorithm. It is easy to verify that this may occur not earlier than n linear equations of the type (37.1) have been imposed because each such an equation eliminates at most one of the nonlinear terms $v_1x, v_2x^2, \ldots, v_nx^n$ from $z(\underline{v},x) - v_0$. To check the latter fact, consider, for instance, the representation of the polynomial $z(\underline{v},x)$ as the inner product of the vector $\underline{X} = [1,x,x^2,\ldots,x^n]$ by the coefficient vector and the influence of the active substitutions on that product. In particular, the substitution (37.1) amounts to summing the vector $[-(f(n,q))^{-1}f(i,q)x^n, i=0,1,\ldots,n]$ with the vector $\underline{X}$ modulo F(x). Similarly at the subsequent elimination steps until $\underline{X}$ turns into the null vector modulo F(x). ■

Remark 37.1. Example 34.1 shows that Theorem 37.1 cannot be applied to λ-algorithms.

Next we will slightly modify the substitution argument following [P64] (compare [BM],pp. 12-15, and [KK]) in order to estimate the number of additions and subtractions in arithmetical algorithms.

Definition 37.1. Let $v_0, v_1, \ldots, v_{N-1}$ be indeterminates. Let $d(0), d(1), \ldots, d(N-1)$ be rational numbers. Then the product $\prod_{i=0}^{N-1} v_i^{d(i)}$ is called a generalized monomial in the indeterminates $v_0, v_1, \ldots, v_{N-1}$. The function $u(v) = \log v$ maps the set of generalized monomials in the N indeterminates $v_0, v_1, \ldots, v_{N-1}$ onto the linear space of N-dimensional vectors with the rational coordinates $d(0), d(1), \ldots, d(N-1)$.

Again we will use the problem of the evaluation of a polynomial to demonstrate the active substitution argument.

Theorem 37.2. Any arithmetical algorithm A for the evaluation of a polynomial $\underline{z}(\underline{v},x)$ of degree n over a field F containing sufficiently many distinct elements involves at least n additions and subtractions.

Remark 37.2. Theorem 37.2 was first proven in [Be] by the algebraic independence argument, which does not work for establishing lower bounds on as(m,n,p).

Proof. We will call the addition or subtraction $R'_q \pm R_q$ passive if $R'_q = 0$, or $R_q = 0$, or both R'_q and R_q lie in F. Otherwise we will call it active. It suffices to consider the case where x=1. Let the algorithm A be represented by the sequence (34.1) and let operation q be the first among the active additions and subtractions. Then in (34.1) each p_g for g<q and therefore also each of R'_q and R_q is a product of an element of F by a generalized monomial in $v_0, v_1, \ldots, v_n$, so that

$$R'_q = f'(q) \prod_{i=0}^{n} v_i^{d'(i)}; \quad R_q = f(q) \prod_{i=0}^{n} v_i^{d(i)}$$

where $f'(q)$ and $f(q)$ are nonzero elements of F; $d'(i)$ and $d(i)$ are rational numbers for all i.

Note that the logarithms of generalized monomials belong to the linear space L(V,F) of linear forms in $v_0,v_1,\ldots,v_n, f (f \in \log F)$ with rational coefficients, so that p_q is the first among the p_g in (34.1) whose logarithm is not in L(V,F). Then impose the following equation on $v_0,v_1,\ldots,v_n$, which turns $\log p_q$ into an element of L(V,F).

$$g R_q' = R_q \tag{37.2}$$

where g is in F. Let $h(i) = d'(i) - d(i)$, $i=0,1,\ldots,n$. Then (37.2) can be equivalently rewritten as follows,

$$g f'(q)(f(q))^{-1} \prod_{i=0}^{n} v_i^{h(i)} = 1. \tag{37.3}$$

In the trivial case where $h(i)=0$ for all i, it is sufficient to choose $g = f(q)(f'(q))^{-1}$ in order to satisfy (37.3). Otherwise we pick any g in F or in an algebraic extension of F such that no divisions by the identical 0 arise in the algorithm A due to imposing (37.2),(37.3). Under (37.2),(37.3), operation q in (34.1) can be replaced by the scalar multiplication

$$p_q = R_q' \pm R_q = R_q'(1 \pm g). \tag{37.4}$$

This reduces the number of additions and subtractions in the algorithm A by 1 because $1 \pm g$ is just a constant from F. Then the same argument is repeated for the first active addition or subtraction that remains in (34.1) under (37.2)-(37.4), and so on until there remain no active additions and subtractions.

We will simplify the rest of the proof assuming that F=C, so that $\log f \in F$ if $f \in F$, $f \neq 0$ (see [BM], pp. 12-15; [KK] for the proof for arbitrary fields F that contain sufficiently many elements). Take logarithms on both sides of (37.3). Then (37.3) is equivalent to the linear equation in $u_i = \log v_i$, $i=0,1,\ldots,n$,

$$\sum_{i=0}^{n} h(i)u_i = \log f(q) - \log g - \log f'(q).$$

Define similar equations at all elimination steps. Note that the elimination process will not stop until those equations turn $z(\underline{v},1) = \sum_{i=0}^{n} v_i$ into a product of an element of F by a generalized monomial. Represent that product as $f2^{L(\underline{u})}$ where $L(\underline{u})$ is a linear function in $\underline{u}$ and $f \in F$. Impose an additional linear equation

$L(\underline{u}) = 0$. That gives a total of $s+1$ linear equations on the u-variables where s is the number of the elimination steps. Therefore $s+1 \geq n+1$ since those linear equations on the u-variables turn $\sum_{i=0}^{n} 2^{u_i}$ into a constant. Thus $s \geq n$. ■

Similarly to the proof of Theorem 37.2 we can obtain that $as(A) \geq (m+p-1)n-1$ for every arithmetical algorithm A for the evaluation of the bilinear form $b(X,Y) = \sum_{j=0}^{n-1} (\sum_{i=0}^{m-1} x_{ij}y_{j0} + \sum_{k=1}^{p-1} x_{0j}y_{jk})$. Therefore already the evaluation of the $m+p-1$ entries of XY, that is, of $\sum_{j=0}^{n-1} x_{ij}y_{j0}$ for $i=0,1,\ldots,m-1$ and $\sum_{j=0}^{n-1} x_{0j}y_{jk}$ for $k=1,2,\ldots,p-1$, requires at least $(m+p-1)(n-1)$ additions and subtractions because summing those $m+p-1$ entries of XY would involve only $m+p-2$ additions and subtractions and would give the bilinear form $b(X,Y)$ as the output.

Finally we note that *our lower estimates for the number* $as(A)$ *of additions and subtractions can be also applied in the case of* λ*-algorithms* A if we consider each λ-algorithm of degree d for the evaluation of $z(\underline{v},x)$ or $\sum_j x_{ij}y_{jk}$ for all i,k as a conventional algorithm modulo λ^{d+1} for the evaluation of $\lambda^d z(\underline{v},x)$ or $\lambda^d \sum_j x_{ij}y_{jk}$ for all i,k, respectively.

Summarizing we have the following result, compare [BM],[KK].

Theorem 37.3. Any arithmetical algorithm or λ-algorithm for mXnXp MM over arbitrary field F of constants containing sufficiently many distinct elements involves at least $(m+p-1)(n-1)$ additions and subtractions.

Remark 37.3. Remark 35.4 can be extended to the case of the lower bounds on $as(A)$.

38. Nonlinear Lower Bounds on the Complexity of Arithmetical Problems Under Additional Restrictions on the Computational Schemes.

Except for cited Strassen's degree method and somewhat similar Ben-Or's (both of them are not efficient for bilinear problems), the attempts of establishing nonlinear lower bounds on the number $ar(A)$ of arithmetical operations in arithmetical algorithms A for arithmetical computational problems have been successful only under some additional strong restrictions on the class of algorithms allowed. Some attempts of that kind followed the idea of [BC] (compare [BM], pp. 121-122) to relate the number $as(A)$ (where A evaluates a set $q_0(V),q_1(V),\ldots,q_{\Gamma-1}(V)$ of quolynomials) to the number of real zeroes of the system of the equations $q_\gamma(V) = 0$, $\gamma = 0,1,\ldots,\Gamma-1$, see [H], [Gri], [Ris] on further study in that direction.

In this section we will consider another interesting approach, due to J. Morgenstern, even though the major application of that approach happened to be not to the general problem of MM but to the problem of the multiplication of a special circulant matrix by a vector, that is, to the following problem of the *discrete Fourier transform* at n points (hereafter referred to as DFT(n)), which has numerous applications in the sciences and technology.

Example 38.1. (DFT(n)). Evaluate the matrix-vector product $\underline{z} = F\underline{v}$ where $\underline{v}$ is an n-dimensional column-vector of indeterminates and F is the following nXn circulant matrix, $F = [\gamma^{ij}, i,j = 0,1,\ldots,n-1]$, γ is a primitive n-th root of 1, such that $\gamma^n = 1$, $\gamma^k \neq 1$ for $0 < k < n$.

Remark 38.1. The famous FFT-algorithm (see, for instance, [AHU],[BM],[K], see also [F72a] for an important interpretation of the FFT via division of polynomials) solves the problem DFT(n) in O(n log n) arithmetical operations as $n \to \infty$. Due to the further improvements presented already in [Ra], [F71], the number of scalar multiplications has been reduced to less than 3n (and then to at most 2n in [W76]) but the current best upper bound on the number of additions and subtractions still remains only O(n log n). Actually the latter algorithms with at most 2n multiplications have quite irregular logical schemes (unlike the FFT-algorithm) and this complicates their implementation. The gap between the upper bound O(n log n) in the FFT-algorithm, and the best current lower bounds of the order n motivates further study.

Since the computational problem DFT(n) is linear, that is, it is a problem of computing a given set of linear functions, such a study can be naturally restricted to the classes of linear algorithms that involve only linear arithmetical operations (additions, subtractions, and scalar multiplications) due to the following simple result that can be derived similarly to Propositions 34.1 and 34.2.

Proposition 38.1 [M73]. If an arithmetical algorithm A computes a set of linear functions, then it is possible to define a linear algorithm L=L(A) that computes that set of functions involving a total of at most 2ar(A) linear operations where at most $\rho(A)$ + as(A) are additions and subtractions.

Note that every bilinear problem can be turned into linear one simply by assigning some constant values to all of the y-variables (or to all of the x-variables) or simply by declaring that all formal power series in the y-variables (or in the x-variables) are considered the constants of the computations. Then Proposition 38.1 can be extended to bilinear problems.

Next we will represent linear algorithms by matrices and then will apply that representation in order to estimate the number of operations involved in those algorithms.

Let a linear algorithm L involve exactly K input indeterminates and ar(L) linear operations. Then an (ar(L) + K) X K matrix M(L) can be associated with L such that the first K rows of M(L) form the K X K identity matrix and the next ar(L) rows are just the coefficient vectors of the outputs of all linear operations of L, provided that such outputs are represented as the linear forms in the K input indeterminates. Let Z designate the submatrix of M(L) formed by the rows associated with the outputs of the problem. (Z depends on the given linear problem but is invariant in the algorithms L.)

We will partition the set of the rows of M(L) into the three subsets: of the first K rows (called input-rows), of the as- rows that represent the outputs of additions and subtractions of L, and of the sm- rows that represent the outputs of scalar multiplications of L. Note that every as-row of M(L) equals the sum or the difference of a pair of some preceding rows of M(L) and every sm-row of M(L) is the product of a preceding row of M(L) by a constant. The latter observation leads to the following result where μ (U) for a matrix U designates maximum |det V| and the maximum is in all square submatrices V of U.

Proposition 38.2 [M73a]. Let L be a linear algorithm over the field of complex numbers C; Z be the matrix associated with the outputs of L; c be the maximum modulus of the constants involved in L. Then

$$(\log c)\ sm(L) + as(L) \geq \log \mu\ (Z).$$

Proof. Let hereafter $M_j(L)$ designate the submatrix of M(L) formed by its first j rows. Let j>K where K is the number of the input-indeterminates of L. Then the above observations about the as-rows and the sm-rows of M(L) (in the paragraph preceding Proposition 38.2) and the elementary properties of the determinant of a matrix imply that $\mu\ (M_j(L)) \leq 2\ \mu\ (M_{j-1}(L))$ if the j-th row of M(L) is an as-row and $\mu\ (M_j(L)) \leq c\ \mu\ (M_{j-1}(L))$ if the j-th row is an sm-row. Thus

$\mu\ (M(L)) \leq c^{sm(L))} 2^{as(L))}\ \mu\ (M_K(L))$. Proposition 38.2 follows from the latter bound because obviously $M_K(L) = I$, $\mu\ (I) = 1$, and $\mu\ (Z) \leq \mu\ (M(L))$ since Z is a submatrix of M(L). ■

The major application of that result is stated in the next corollary.

Corollary 38.3 [M73a]. Every linear algorithm L for DFT(n), see Example 38.1, involves at least 0.5 n log n arithmetical operations provided that the modulus of no scalar involved in L exceeds 2.

Proof. Note that F/||F|| is a unitary matrix where F is the output matrix of the problem DFT(n). Note that $||F|| = n^{1/2}$. Apply Proposition 29.2 and obtain that $|\det F| = n^{n/2}$. Then apply Proposition 38.2 where Z=F. ■

Morgenstern's approach can be applied in order to relate the numbers of additions and subtractions required for MM (and for other bilinear computational problems) with the rank of tensors of some associated polylinear forms, see [P81b].

On the other hand, Corollary 38.3 might help designers of fast algorithms for DFT(n) in narrowing the domain of their search. Similar applications can be expected of the nonlinear lower bounds on the arithmetical complexity of linear and bilinear problems derived in [HS], [V], and [P83] under certain restrictions on the class of allowed computational schemes. In the next section we will derive the bounds of [P83]. Our analysis will lead us a certain quantity for measuring the asynchronicity of linear and bilinear algorithms. Defining such a quantity will actually be the main objective of our study in the next section.

39. A Trade-off between the Additive Complexity and the Asynchronicity of Linear and Bilinear Algorithms.

To fulfill our promise given at the very end of the previous section, we will start with seeking for the lower bounds on the number nas(L) of nonscalar additions and subtractions in linear algorithms L that computes a given set of linear functions in K indeterminates. We will count only nonscalar additive operations (that is, such additions and substractions that involve no constants) and consider other operations cost-free, so that all of the K inputs and of the nas(L) computed values are defined only up to within their multiplications by nonzero constants and the additions of constants to them.

We will reduce the study of such an algorithm L to the study of the following associated acyclic directed graph $G=(V,E)$. The vertex set V of G is the set of all of the K inputs of L and of the nas(L) outputs of nonscalar additions and subtractions. The edge set E is the set of all of the nas(L) non-scalar additions and subtractions of L. The incidence relation $e = (u \rightarrow v)$ where u and v are in V and e is in E holds if and only if e is a nonscalar operation in L with the output v and with u being one of the two inputs of that operation. Then the indegree of every vertex v of G is equal either to 0 if the vertex v is an input-indeterminate of L or to 2 otherwise. The outdegrees of the vertices may equal 0 only for (some of) the outputs of L.

The contribution of an input s of L to an output t of L can be seen by tracing the directed paths from s to t in G. Studying the structure of the algorithm L, we will use the following definition and notation.

Definition 39.1. Let all edges of G have the same cost 1. Let S designate the set of all input-vertices (of indegree 0). Let S(v) designate the set of all predecessors (ancestors) of a vertex v in S. Let $\sigma(u,v)$ and $\lambda(u,v)$ designate the costs of a shortest path and of a longest path from a vertex u to a vertex v, respectively. Let

$$\sigma(v) = \underset{s \in S(v)}{\text{minimum}}\ \sigma(s,v),\ \lambda(v) = \underset{s \in S(v)}{\text{maximum}}\ \lambda(s,v),\ \lambda = \underset{v \in V}{\text{maximum}}\ \lambda(v). \qquad (39.1)$$

In the sequel we will use the partition of the vertex set $V = \bigcup_j V_j$ into the distance sets V_j (layers) where V_j consists of vertices v such that $\lambda(v) = j$, $j=0,1,\ldots,\lambda$; note that $V_0 = S$.

We will consider the case where the linear computation by L is performed for the problem where

$$|S(t)| = n \text{ for all outputs } t, \qquad (39.2)$$

that is, where every output depends on the same number n of indeterminates. (39.2) holds for DFT(n) and for $n \times n$ MM where in the latter case all y-variables or all x-variables are declared to be constants (this turns MM into a linear problem).

Since the indegrees of all vertices are equal to 0 or 2, we arrive at the following fact.

<u>Proposition</u> <u>39.1</u>. Let (39.2) hold. Then for every output vertex t we have that $t \in V_j$ for $j \geq \log n$, that is, the maximum cost of the longest path from s to t is not less than log n. Here the maximum is in all $s \in S(t)$.

Next we will define the class of linear algorithms L that may immediately serve as the schemes of the computation in parallel. (This will be only an auxiliary potential property of those algorithms; we will not study them as parallel algorithms in this book.)

<u>Definition</u> <u>39.2</u>. <u>A linear algorithm is synchronized</u> if for all output vertices t of the associated digraph their shortest and longest distances to the input set S equal λ, that is,

$$\sigma(t) = \lambda(t) = \lambda \text{ for all output vertices } t, \qquad (39.3)$$

compare (39.1). <u>A bilinear algorithm</u> for a bilinear problem $\{b_y(X,Y)\}$ <u>is</u> <u>x-synchronized</u> (<u>or</u> y-<u>synchronized</u>) if it is turned into a synchronized linear algorithm by the declaration that all y-variables are constants (or that all x-variables are constants, respectively).

The partition of the associated digraph G of a synchronized algorithm L into the distance sets V_j is structured as follows. Every vertex of the distance set V_{j+1} for a natural j is a linear combination of its two immediate predecessors, both in the set V_j. (Otherwise (39.3) will not hold.) This implies the following result.

<u>Proposition</u> <u>39.2</u>. Let G be the digraph associated with a synchronized algorithm L. Then, for all natural i and j such that $i+j \leq \lambda$, every vertex of the set V_{i+j} is a linear combination of the vertices of the set V_i.

Next we will deduce the following result from Propositions 39.1 and 39.2.

<u>Proposition</u> <u>39.3</u>. Let L be a synchronized linear algorithm for the computation of Q linearly independent functions, each of them depending on at least n input-indeterminates (compare (39.2)). Then L involves at least Q log n nonscalar additions and subtractions.

Before proving Proposition 39.3, we will state its two immediate corollaries.

<u>Corollary</u> <u>39.4</u>. Every synchronized linear algorithm for DFT(n) involves at least n log n nonscalar additions and subtractions.

<u>Corollary</u> <u>39.5</u>. Every synchronized bilinear algorithm for $n \times n$ MM involves at least $n^2 \log n$ nonscalar additions and subtractions.

<u>Remark</u> <u>39.1</u>. Similar results can be obtained for the problems of convolution and cyclic convolution, defined in Example 2.2.

Proof of Proposition 39.3. Since every vertex of indegree 2 is associated with a nonscalar addition or subtraction, Proposition 39.3 will follow if we prove that the cardinality of the vertex set V is at least $Q \log n + K$ where $K = |S|$ is the number of input-vertices of indegree 0. It follows from Proposition 39.1 that the set V consists of at least $1 + \log n$ different distance sets V_j, $j=0,1,\dots,\lambda$. It remains to deduce from (39.3) that the cardinality of each distance set is at least Q, $|V_j| \geq Q$ for all j. The latter property follows because for every natural j the Q linearly independent outputs are linear combinations of the vertices of V_j, see Proposition 39.2. ∎

Next we will introduce a quantitative measure for the amount of asynchronicity of a linear algorithm (that terminology is due to the suggestion by S. Cook).

Transform the digraph G associated with a given (asynchronized) algorithm as follows. Label every edge $e = (u \to v)$ of G by $h = \lambda(v) - \lambda(u) - 1$, compare (39.1). (Such a value $h=h(e)$ will be called the asynchronicity of an edge e.) Then replace the edge e by the directed path $u \to u(1) \to \dots \to u(h) \to v$, that is, substitute the sequence of the h+1 edges $e(0),\dots,e(h)$, where $e(i) = (u(i) \to u(i+1))$, $i=0,1,\dots,h$, $u(0)=u, u(h+1)=v$, for the edge e in E and include the h new vertices $u(1),\dots,u(h)$ into V. (This can be interpreted as the h step delay of the $\pm$ operation associated with e and as the storage of its input associated with u.) The new vertices will have indegrees and outdegrees 1; their longest distances from the input set S will be as follows, $\lambda(u(k)) = \lambda(u)+k$. In the result of such a transformation applied to all edges of the original digraph G where $h>0$, we will arrive at a new digraph $G^* = (V^*,E^*)$ that satisfies the property (39.3). We will also have the following relation between the cardinalities of V and V^*,

$$|V^*| = |V| + H, \quad H = \sum_{e \in E} h(e). \tag{39.4}$$

We will call H the asynchronicity measure of the associated linear algorithm.

Now we may extend Proposition 39.3 as follows.

Theorem 39.6. Let L be a linear algorithm for computing Q linearly independent functions, each of them depending on at least n input-indeterminates. Then L involves at least $Q \log n - H$ arithmetical operations where H is the asynchronicity measure of L defined by (39.4).

It is easy to obtain similar extensions of Corollaries 39.4 and 39.5 and of Remark 39.1. We leave this to the reader.

The latter theorem indicates the necessary amount of asynchronicity in the algorithms whose speed would exceed the barrier $Q \log n$. That theorem was also an occasion for us to motivate the introduction of some quantitative measure of the asynchronicity that might be useful in the combinatorial study of parallel algorithms.

40. An Attempt of Practical Acceleration of Matrix Multiplication and of Some Other Arithmetical Computations.

In this section we will consider the following generalization of the class of λ-algorithms for the evaluation of a set of rational expressions (quolynomials), which promises to be practically efficient, see Examples 40.1-40.5 below and our comments that follow those examples.

Definition 40.1. Let the sequence (34.1) satisfy all requirements of Definition 34.2 except that each of R_s and R_s' is allowed to be equal to a constant, to an input indeterminate, or to any coefficient of a polynomial p_q in λ for any $q<s$; also let the following weaker requirement substitute for (34.5): there exist nonnegative integers $d(g)$ and $s(g)$ such that $s(g) < S$, $p_{s(g)}(\lambda)$ are polynomials in λ, and $Q_g(V)$ are equal to the $\lambda^{d(g)}$-coefficients of those polynomials $p_{s(g)}(\lambda)$ for $g=0,1,\ldots,G-1$. Then the sequence (34.1) is called a straight line arithmetical unrestricted λ-algorithm of degree $d = \underset{g}{\text{maximum}}\, d(g)$ for the evaluation of a given set of rational functions (quolynomials) $\{Q_g(V),\ g=0,1,\ldots,G-1\}$ over the given ring F.

In many cases we may substantially simplify the computation using the unrestricted λ-algorithms. Here are some examples of the unrestricted λ-algorithms where the computation has been reduced to a single multiplication, not counting the complexity of representing the inputs as polynomials in λ and of recovering the coefficients of the polynomials in λ.

Example 40.1, the multiplication of two complex numbers, compare Example 2.1.

$$p_0 = (x_0 + \lambda x_1)(\lambda y_0 - y_1 + \lambda^2 y_1).$$

The values $z_0 = x_0y_0 - x_1y_1$ and $z_1 = x_0y_1 + x_1y_0$ are recovered as the λ-coefficient of p_0 and the λ^2-coefficient of p_0, respectively. Here the degree d of the unrestricted λ-algorithm is 2.

The next example is straightforward but we will include it for the sake of completeness.

Example 40.2, polynomial multiplication, compare Examples 2.2 and 18.1.

$$p_0 = \left(\sum_{i=0}^{m-1} \lambda^i x_i\right)\left(\sum_{j=0}^{n-1} \lambda^j y_j\right).$$

The desired values $z_k = \sum_{i=0}^{k} x_i y_{k-i}$, $k=0,1,\ldots,m+n-2$ (where $x_i = y_j = 0$ for $i>m-1$, $j<0$, and $j>n-1$) are equal to the λ^k-coefficients of p_0. The degree d of that unrestricted λ-algorithm is $m+n-2$.

Example 40.3, the evaluation of the inner (scalar) product of two vectors, $z = \underline{x}^T\underline{y} = \sum_{j=0}^{n-1} x_{0j}y_{j0}$, that is, 1XnX1 MM.

$$p_0 = (\sum_{g=0}^{n-1} \lambda^g x_{0g}) (\sum_{h=0}^{n-1} \lambda^{n-1-h} y_{h0}).$$

The scalar product $z = \underline{x}^T\underline{y}$ is equal to the λ^{n-1} coefficient of p_0. The degree d of this unrestricted λ-algorithm is n-1.

Example 40.4, the evaluation of the outer product of two vectors $\{z_{ik} = x_{i0}y_{0k},\ i=0,1,\ldots,m-1;\ k=0,1,\ldots,p-1\}$, that is, mX1Xp MM.

$$p_0 = (\sum_{i=0}^{m-1} \lambda^{ip} x_{i0}) (\sum_{k=0}^{p-1} \lambda^k y_{0k}).$$

z_{ik} is equal to the λ^{ip+k}-coefficient of p_0 for each pair i,k. The degree d=mp-1 in this case.

Example 40.5, mXnXp MM, that is, the evaluation of $z_{ik} = \sum_{j=0}^{n-1} x_{ij}y_{jk}$ for $i=0,1,\ldots,m-1;\ k=0,1,\ldots,p-1$, compare Examples 40.3 and 40.4.

$$p_0 = (\sum_{i=0}^{m-1}\sum_{g=0}^{n-1} \lambda^{(2n-1)ip+g} x_{ig}) (\sum_{h=0}^{n-1}\sum_{k=0}^{p-1} \lambda^{(2n-1)k+n-1-h} y_{hk}).$$

The value z_{ik} is equal to the $\lambda^{d(i,k)}$-coefficient of p_0 for each pair i,k where $d(i,k)=(2n-1)(ip+k)+n-1$. The degree d of such an unrestricted λ-algorithm is equal to $(2n-1)(mp-1)+n-1$.

The unrestricted λ-algorithms of Examples 40.1-40.5 can be incorporated into the computation for other problems. For instance, Gaussian elimination for SLE, Det, and MI can be performed using the unrestricted λ-algorithm of Example 40.4 at every elimination step. The iterative algorithms for SLE may include the unrestricted λ-algorithm of Example 40.5 (for p=1) at every iteration, which can be also used for the QR-factorization of a matrix.

The unrestricted λ-algorithms reduce the evaluation over a given field or ring of constants F to the evaluation over a ring of polynomials in λ over F. In Examples 40.2-40.5 all outputs are natural numbers if the inputs are natural numbers. The upper bounds on the magnitudes of the inputs define some upper bounds on the magnitudes of the outputs. Suppose that all outputs are known to be natural numbers that are less than K and that the degree of the unrestricted λ-algorithm is d. Then we may substitute $\lambda = 2^G$, $G = \lceil \log K \rceil$ into the λ-algorithm, perform all computations modulo $2^{G(d+1)}$ (compare Example 18.1 and [P82]), and finally recover the desired output values from the computed binary numbers. The extension from the case of natural inputs to the cases of integer inputs and finite binary inputs can be done as in Example 18.1.

Actually the reduction of the number of arithmetical operations has been achieved only by the price of increasing the precision as in Section 18, see Example 18.1 and Yuval's algorithm for the shortest paths. However, if such a computation is implemented on the computers capable to deal with our "long" binary numbers as with single precision numbers, then unrestricted λ-algorithms can be highly recommended.

If necessary, the positive degree d can be always reduced by involving more arithmetical operations. For instance, we may reduce the 1XNX1 MM, N=3s, to s applications of the unrestricted λ-algorithm of Example 40.3 where n=3 and d=2. This will still save about 2/3 operations comparing with the best conventional algorithm for that problem. Similarly we may partition mXnXp MM into subproblems that can be efficiently solved by unrestricted λ-algorithms of smaller degrees.

Remark 40.1. One may try to devise unrestricted bilinear λ-algorithms of relatively small degrees d(n) and of ranks M(n) for nXn MM such that, say, $\log(M(n)(2d(n)+1))/\log n < 2.495$. This would imply the reduction of the exponent of MM at least over infinite fields because Proposition 6.1 can be extended to the case of unrestricted λ-algorithms. This approach does not seem promising, however. Furthermore, even in the case of its success, the resulting algorithms will be superior over the straightforward method only for enormously large problems and will surely have no practical value (similarly to the algorithms of [CW]). It seems more promising to seek for fast efficient methods for computing the selected segments of the products of two long integers where those integers represent λ-polynomials whose coefficients are the integer inputs of the problem of MM, where $\lambda = 2^G$, where G is a sufficiently large natural number, and where the selected segments define the outputs of the problems of Examples 40.3-40.5.

Remark 40.2. This section continues our study begun in [P80a] and [P82].

APPENDIX (An Abstract of a Result at Oberwalfach 1979)

This appendix contains a copy of the last page of the collection of abstracts presented at the international complexity conference in Oberwolfach (West Germany) in October 1979. In spite of his signature under the abstract, S. Winograd insisted (in writing to the present author) that he should not be credited for the exponent 2.522 because his participation was not original. Indeed, it was limited only to reconstruction of a theorem (see Theorem 7.2 in Section 7) after its sketchy presentation by A. Schönhage at that conference. However, the reconstruction did require substantial work since Schönhage's presentation of that theorem was indeed quite sketchy.

Complexity of Systems of
Algebraic Equations
Daniel Lazard

Let $f_1, \ldots, f_k$ be k polynomials in n indeterminates which have a finite number of common zeros in the algebraic closure of the ground field, counting the zeros at infini[ty]. An algorithm is given which computes all those zeros. If d is the ~~degree~~ highest degree of the polynomials, the computations needed by this algorithm consist in the resolution of one univariate polynomial whose degree is the number of solutions and a number of operations of the ground field which is polynomial in $(ed)^n$.

As of 21.24 hr. of October 26, 1979
the best known ~~example~~ exponent for
matrix multiplication is 2.521812716

VICTOR PAN & SHMUEL WINOGRAD

Index of Some Concepts

REFERENCES

[A] K. Atkinson 1978, An Introduction to Numerical Analysis, Wiley, N.Y.

[AHU] A.V. Aho, J.E. Hopcroft, J.D. Ullman 1976, The Design and Analysis of Computer Algorithms, Addison-Wesley, Mass.

[AS] A. Alder and V. Strassen 1981, On the Algorithmic Complexity of the Associative Algebras, Theoretical Computer Science, 15, 201-211.

[B80] D. Bini 1980, Relations Between EC-algorithms and APA-algorithms. Applications, Calcolo, XVII, 87-97.

[B82] D. Bini 1982, Reply to the Paper "The Numerical Instability of Bini's Algorithm", Information Processing Letters, 14,3, 144-145.

[B83] D. Bini 1983, On a Class of Matrices Related to Toeplitz Matrices, Technical Report, TR 83-5, Computer Science Department, SUNY Albany, N.Y.

[B84] D. Bini 1984, On Commutativity and Approximation, Theoretical Computer Science, 28, 1-2, 135-150.

[BC] A. Borodin and S. Cook 1974, On the Number of Additions to Compute Specific Polynomials, Proc. Sixth Ann. ACM Symp. on Theory of Computing, 342-347.

[BCLR] D. Bini, M. Capovani, G. Lotti, F. Romani 1979, $O(n^{2.7799})$ Complexity for Matrix Multiplication, Information Processing Letters, 8,5, 234-235.

[BD76] R.W. Brockett and D. Dobkin 1976, On the Number of Multiplications Required for Matrix Multiplication, SIAM J. Computing, 5,4, 624-628.

[BD78] R.W. Brockett and D. Dobkin 1978, On the Optimal Evaluation of a Set of Bilinear Forms, Linear Algebra and Its Applications, 19, 207-235.

[Be] E.G. Belaga 1958, Some Problems in the Computation of Polynomials, Doklady Academii Nauk SSSR, 123, 775-777 (in Russian).

[BGH] A. Borodin, I. von zur Gathen, J. Hopcroft 1982, Fast Parallel Matrix and GCD Computations, Proc. Twenty Third Ann. Symp. on Foundations of Computer Science, 65-71.

[BGY] R.P. Brent, F.G. Gustavson, D.Y.Y. Yun 1980, Fast Solution of Toeplitz System of Equations and Computation of Padé Approximants, J. of Algorithms, 1, 259-295.

[Bl] P. Bloniarz 1980, A Shortest Path Algorithm with Expected Time $O(n^2 \log n \log^* n)$, Proc. Twelfth Ann. ACM Symposium on Theory of Computing, 378-384.

[BM] A. Borodin and I. Munro 1975, The Computational Complexity of Algeraic and Numeric Problems, American Elsevier, N.Y.

[BML]G. Birkhoff and S. MacLane 1953, A Survey of Modern Algebra, MacMillan, N.Y.

[BO] M. Ben-Or 1983, Lower Bounds for Algebraic Computation Tree, Proc. Fifteenth Ann. ACM Symp. on Theory of Computing, 80-86.

[Bou]N. Bourbaki 1970, Algèbre 2, Hermann, Paris.

[Br70]R.P. Brent 1970, Error Analysis of Algorithms for Matrix Multiplication and Triangular Decompositions Using Winograd's Identity, Numerische Math., 16, 145-156.

[Br76]R.P. Brent 1976, Multiple-Precision Zero-Finding Methods and the Complexity of Elementary Function Evaluation, in Analytic Computational Complexity (J.F. Traub ed.), Academic Press, N.Y., 151-176.

[BS] W. Baur and V. Strassen 1983, On the Complexity of Partial Derivatives, Theoretical Computer Science, 22, 317-330.

[C] D. Coppersmith 1982, Rapid Multiplication of Rectangular Matrices, SIAM J. on Computing, 11,3, 467-471.

[CdB]S.D. Conte and C. de Boor 1980, Elementary Numerical Analysis (an Algorithmic Approach), McGraw-Hill, N.Y.

[CW] D. Coppersmith and S. Winograd 1980, On the Asymptotic Complexity of Matrix Multiplication, SIAM J. on Computing, 11,3, 472-492.

[DB] G. Dahlquist and A. Björk 1974, Numerical Methods, Prentice-Hall, N.J.

[dG] H.F. de Groote 1978, On Varieties of Optimal Algorithms for the Computation of Bilinear Mappings, Theoretical Computer Science, 7, 1-24, 127-148.

[DGK]J.J. Dongarra, F.G. Gustavson, A. Karp 1984, Implementing Linear Algebra Algorithms for Dense Matrices on a Vector Pipeline Machine, SIAM Review, 26,1, 91-112.

[F71]C.M. Fiduccia 1971, Fast Matrix Multiplication, Proc. Third Ann. ACM Symp. on Theory of Computing, 45-49.

[F72]C.M. Fiduccia 1972, On Obtaining Upper Bounds on the Complexity of Matrix Multiplication, in Complexity of Computer Computations, (R.E. Miller and J.W. Thatcher, eds.) Plenum Press, N.Y., 31-40.

[F72a]C.M. Fiduccia 1972, Polynomial Evaluation via the Division Algorithms. The Fast Fourier Transform Revisited. Proc. Fourth Ann. ACM Symp. on Theory of Computing, 88-93.

[FF] D.K. Faddeev and V.N. Faddeeva 1963, Computational Methods of Linear Algebra, W.H. Freeman, San Francisco, Calif.

[FM] M.J. Fisher and A.R. Meyer 1971, Boolean Matrix Multiplication and Transitive Closure, Conference Record, IEEE Twelfth Ann. Symp. on Switching and Automata Theory, 129-131.

[FMKL]B. Friedlander, M. Morf, T. Kailath, L. Ljung 1979, New Inversion Formulas for Matrices Classified in Terms of Their Distances from Toeplitz Matrices, Linear Algebra and Its Applications, 27, 31-60.

[Fr] M.L. Fredman 1976, New Bounds on the Complexity of the Shortest Path Problem, SIAM J. on Computing, 5, 83-89.

[FZ] C.M. Fiduccia and Y. Zalstein 1975, Algebras Having Linear Multiplicative Complexity, J. ACM, 24,2, 311-331.

[Gre]R.T. Gregory 1980, Error-Free Computation: Why It Is Needed and Methods for Doing It, R.E. Krieger, N.Y.

[Gri]D. Yu. Grigoriev 1982, Lower Bounds in the Algebraic Computational Complexity, in The Theory of the Complexity of Computations (D. Yu. Grigoriev and A.O. Slisenko eds.), Nauka, Leningrad (in Russian).

[GvL]G.H. Golub and C.F. van Loan 1983, Matrix Computations, Johns Hopkins University Press, Baltimore, MD.

[H] A.G. Hovansky 1980, On a Class of Systems of Transcendental Equations, Doklady Akademii Nauk SSSR, 244,5, 1072-1076 (in Russian), Transl. in Soviet Math. Dokl., 22,3, 762-765.

[He] D. Heller 1978, A Survey of Parallel Algorithms in Numerical Linear Algebra, SIAM Review, 20, 740-777.

[HK69]J.E. Hopcroft and L.R. Kerr 1969, Some Techniques for Proving Certain Simple Programs Optimal, Proc. Tenth Ann. Symp. on Switching and Automata Theory, 36-45.

[HK71]J.E. Hopcroft and L.R. Kerr 1971, On Minimizing the Number of Multiplications Necessary for Matrix Multiplication, SIAM J. Applied Math., 20, 30-36.

[HM] J.E. Hopcroft and J. Musinski 1973, Duality Applied to the Complexity of Matrix Multiplication and Other Bilinear Forms, SIAM J. on Computing, 2,3, 159-173.

[How]T.D. Howell 1976, Tensor Rank and the Complexity of Bilinear Forms, Ph.D. Thesis, Dept. of Computer Sci., Cornell Univ.

[HS] L.H. Harper and J.E. Savage 1979, Lower Bounds on Synchronous Combinatorial Complexity, SIAM J. on Computing, 8,2, 115-119.

[IMH]O.H. Ibarra, S. Moran, R. Hui 1982, A Generalization of the Fast LUP Matrix Decomposition Algorithm and Applications, J. of Algorithms, 3, 45-56.

[JJT]J. Ja' Ja' and J. Takche 1983, Improved Lower Bounds for Some Matrix Multiplication Problems, Tech. Rep. Dept. Electrical Engineering, Univ. of Maryland, College Park, MD (submitted to Information Processing Letters).

[K] D.E. Knuth 1981, The Art of Computer Programming: Seminumerical Algorithms, 2, Addison-Wesley, Mass.

[Ke] W. Keller 1982, Fast Algorithms for Characteristic Polynomial, Tech. Rep., Institut fur Angewandte Mathematic, Universitat Zurich, Switzerland.

[KK] Z.M. Kedem and D.G. Kirkpatrick 1977, Addition Requirements for Rational Functions, SIAM J. on Computing, 2,3, 188-199.

[KKM] T. Kailath, S.-Y. Kung, M. Morf 1979, Displacement Ranks of Matrices and Linear Equations, J. of Math. Analysis and Applics., 68, 395-407.

[L] J.D. Laderman 1976, A Noncommutative Algorithm for Multiplying 3X3 Matrices Using 23 Multiplications, Bull. Amer. Math. Soc., 82, 126-128.

[La] E.L. Lawler 1976, Combinatorial Optimization: Networks and Matroids, Holt, Rinehart and Winston, N.Y.

[LW] J.C. Lafon and S. Winograd 1979, A Lower Bound for the Multiplicative Complexity of the Product of Two Matrices, Tech. Rep., Seminaire d'Informatique, Centre de Calcul de l'Esplanade, Université Louis Pasteur, Strasbourg, France.

[M73] J. Morgenstern 1973, Algorithmes Lineaires Tangents et Complexité, C.R. acad. Sc. Paris, 277, Serie A, 367-369.

[M73a] J. Morgenstern 1973, Note on a Lower Bound of the Linear Complexity of the Fast Fourier Transform, J. ACM, 20,2, 305-306.

[Mi] W. Miller 1972, On the Stability of Finite Numerical Procedures, Numerische Math., 19, 425-432.

[Mo] T.S. Motzkin 1955, Evaluation of Polynomials and Evaluation of Rational Functions, Bull. Amer. Math. Soc., 61, 9, 163.

[MP] W.L. Miranker and V.Ya. Pan 1980, Methods of Aggregations, Linear Algebra and Its Applications, 29, 231-257.

[Of] Yu. Ofman 1963, On Algorithmic Complexity of Discrete Functions, Doklady Akademii Nauk SSSR, 145,1 (1963), 48-51 (in Russian), transl. in Soviet Physics Dokl., 7,7, 589-591.

[Os] A.M. Ostrowski 1954, On Two Problems in Abstract Algebra Connected with Horner's Rule, Studies presented to R. von Mises, Academic Press, N.Y., 40-48.

[P62] V.Ya. Pan 1962, On Some Methods of Computing Polynomial Values, Problemy Kibernetiki, 7, 21-30 (in Russian), transl. in Problems of Cybernetics, A.A. Lyapunov edit., USSR, 7, 20-30, U.S. Dept. of Commerce.

[P64] V.Ya. Pan 1964, Methods for Computing Polynomials, Ph.D. Thesis, Dept. of Mechanics and Mathematics, Moscow University (in Russian).

[P66] V.Ya. Pan 1966, On Methods of Computing Polynomial Values, Uspekhi Matematicheskih Nauk, 21,1(127), 103-134 (in Russian), transl. in Russian Mathematical Surveys, 21,1(127), 105-137.

[P72]V.Ya. Pan 1972, On Schemes for the Computation of Products and Inverses of Matrices, Uspekhi Matematicheskih Nauk, 27,5(167), 249-250 (in Russian).

[P78]V.Ya. Pan 1978, Strassen Algorithm Is Not Optimal. Trilinear Technique of Aggregating, Uniting and Canceling for Constructing Fast Algorithms for Matrix Multiplication, Proc. Nineteenth Ann. Symp. on Foundations of Computer Science, 166-176.

[P79]V.Ya. Pan 1979, Field Extension and Trilinear Aggregating, Uniting and Cancelling for the Acceleration of Matrix Multiplication, Proc. Twentieth Ann. Symp. on Foundations of Computer Science, 28-38.

[P80]V.Ya. Pan 1980, New Fast Algorithms for Matrix Operations, SIAM J. on Computing, 9,2, 321-342.

[P80a]V.Ya. Pan 1980, The Bit-Operation Complexity of the Convolution of Vectors and of the DFT, Tech. Rep. 80-6, Computer Science Dept., SUNY Albany, Albany, N.Y.

[P80b]V.Ya. Pan 1980, Less Than 2.5161 Exponent for Matrix Multiplication, Abstracts Presented to Amer. Math. Soc., 1,4, 384.

[P81]V.Ya. Pan 1981, New Combinations of Methods for the Acceleration of Matrix Multiplication, Computers and Math. (with Applics.), 7,1, 73-125.

[P81a]V.Ya. Pan 1981, The Bit-Operation Complexity of Matrix Multiplication and of All Pair Shortest Path Problem, Computers and Math. (with Applics.), 7,5, 431-438.

[P81b]V.Ya. Pan 1981, The Lower Bound on the Additive Complexity of Bilinear Problems in Terms of Some Algebraic Quantities, Information Processing Letters, 13,2, 71-72.

[P81c]V.Ya. Pan 1981, The Bit-Complexity of Arithmetic Algorithms, J. of Algorithms, 1, 144-163.

[P81d]V.Ya. Pan 1981, A Unified Approach to the Analysis of Bilinear Algorithms, J. of Algorithms, 2, 301-310.

[P82]V.Ya. Pan 1982, The Bit-Operation Complexity of Approximate Evaluation of Matrix and Polynomial Products Using Modular Arithmetic, Computers and Math. (with Applics.), 8,2, 137-140.

[P82a]V.Ya. Pan 1982, Trilinear Aggregating with Implicit Canceling for a New Acceleration of Matrix Multiplication, Computers and Math. (with Applics.), 8,1, 23-34.

[P82b]V.Ya. Pan 1982, Fast Matrix Multiplication without APA-Algorithms, Computers and Math. (with Applics.), 8,5, 343-366.

[P82c]V.Ya. Pan 1982, Trilinear Aggregating is the Basis for the Asymptotically Fastest Known Algorithms for Matrix Multiplication, Conference Record, Second Conference on Foundations of Software Technology and Theoretical Computer Science, Bangalore, India, 321-337.

[P83]V.Ya. Pan 1983, The Additive and Logical Complexities of Linear and Bilinear Algorithms, J. of Algorithms, 4, 1-34.

[P84]V.Ya. Pan 1984, How Can We Speed Up Matrix Multiplication?, SIAM Review, 26,3, 393-415.

[P84a]V.Ya. Pan 1984, Trilinear Aggregating and the Recent Asymptotic Acceleration of Matrix Multiplication, to appear in Theoretical Computer Science.

[Pr73]R.L. Probert 1973, On the Complexity of Matrix Multiplication, Tech. Rep. CS-73-27, Computer Science Dept., University of Waterloo, Canada.

[Pr76]R.L. Probert 1976, On the Composition of Matrix Multiplication Algorithms, Proc. Sixth Manitoba Conference on Numerical Math., 357-366.

[R80]F. Romani 1980, Complexity Measures for Matrix Multiplication Algorithms, Calcolo, 17,1, 77-86.

[R80a]F. Romani 1980, Shortest Path Problem Is Not Harder Than Matrix Multiplication, Information Processing Letters, 11,3, 134-136.

[R82]F. Romani 1982, Some Properties of Disjoint Sums of Tensors Related to MM, SIAM J. Computing, 11,2, 263-267.

[Ra] C.M. Rader 1968, Discrete Fourier Transform When the Number of Data Samples Is Prime, Proc. IEEE, 56, 1107-1108.

[Ric]J.R. Rice 1983, Numerical Methods, Software and Analysis, McGraw Hill, N.Y.

[Ris]J.J. Risler, Additive Complexity and Zeroes of Real Polynomials, to appear in SIAM J. on Computing.

[S73]A. Schönhage 1973, Fast Schmidt Orthogonalization and Unitary Transformations of Large Matrices, in Complexity of Sequential and Parallel Numerical Algorithms (J.F. Traub, edit.), Academic Press, N.Y.

[S79]A. Schönhage 1979, Partial and Total Matrix Multiplication, Tech. Report, Universität Tübingen.

[S80]A. Schönhage 1980, private communication.

[S80a]A. Schönhage 1980, Storage Modification Machines, SIAM J. on Computing, 9, 490-508.

[S81]A. Schönhage 1981, Partial and Total Matrix Multiplication, SIAM J. on Computing, 10,3, 434-456.

[Scha]G. Schachtel 1978, A Non-commutative Algorithm for Multiplying 5X5 Matrices Using 103 Multiplications, Information Processing Letters, 4, 180-182.

[Schn]C.P. Schnorr 1981, An Extension of Strassen's Degree Bound, SIAM J. on Computing, 10, 371-382.

[SS] A. Schönhage and V. Strassen 1971, Schnelle Multiplikation Grosser Zahlen, Computing, 7, 281-292.

[St69]V. Strassen 1969, Gaussian Elimination Is Not Optimal, Numerische Math., 13, 354-356.

[St72]V. Strassen 1972, Evaluation of Rational Functions, in Complexity of Computer Computations, (R.E. Miller and J. W. Thatcher edits.), Plenum Press, N.Y., 1-10.

[St73]V. Strassen 1973, Vermeidung von Division, J. Reine Angew. Math., 264, 184-202.

[St73a]V. Strassen 1973, Die Berechungskomplexität von elementarsymmetrischen Functionen und von Interpolationskoeffizienten, Numerische Math., 20, 238-251.

[St81]V. Strassen 1981, The Computational Complexity of Continued Fractions, Proc. ACM Symp. on Symbolic and Algebraic Computation, 51-67.

[T] A.L. Toom 1963, The Complexity of a Scheme of Functional Elements Realizing the Multiplication of Integers, Doklady Academii Nauk SSSR, 150, 496-498 (in Russian), transl. in Soviet Mathematics, 3, 714-716.

[Ta] R.E. Tarjan 1983, Data Structure and Network Algorithms, CBMS-NSF Regional Conference Series in Applied Mathematics, SIAM, Philadelphia, PA.

[Tr] J.F. Traub (edit.) 1976, Analitic Computational Complexity, Academic Press, N.Y.

[TW] J.F. Traub and H. Wozniakowski 1980, A Generalized Theory of Optimal Algorithms, Academic Press, New York.

[V] L.G. Valiant 1975, On Nonlinear Lower Bounds in Computational Complexity, Proc. Seventh Ann. ACM Symp. on Theory of Computing, 45-53.

[vdW]B.L. van der Waerden 1970, Algebra, vv.1,2, Ungar, N.Y.

[W67]S. Winograd 1967, On the Number of Multiplications Required to Compute Certain Functions, Proc. National Academy of Science, 58, 1840-1842.

[W68]S. Winograd 1968, A New Algorithm for Inner Product, IEEE Trans. on Computers, C-17, 693-694.

[W70]S. Winograd 1970, On the Number of Multiplications Necessary to Compute Certain Functions, Comm. on Pure and Applied Math., 23, 165-179.

[W71]S. Winograd 1971, On Multiplication of 2X2 Matrices, Linear Algebra and Its Applications, 4, 381-388.

[W78]S. Winograd 1978, On Computing the Discrete Fourier Transform, Math. Computation, 32, 175-199.

[W80]S. Winograd 1980, Arithmetic Complexity of Computations, CBMS-NSF Regional Conference Series in Applied Math., SIAM, Philadelphia, PA.

[Wa] A. Waksman 1970, On Winograd's Algorithm for Inner Product, IEEE Trans. Computers, C-119, 360-361.

[Wil]J.H. Wilkinson 1965, The Algebraic Eigenvalue Problem, Clarendon Press, Oxford.

[Y] G. Yuval 1976, An Algorithm for Finding All Shortest Paths Using $N^{2.81}$ Infinite Precision Multiplications, Information Processing Letters, 4,6, 155-156.